D0948788

Tomato Growing

Tomato Growing

A Programme for
Successful Cultivation
Under Glass

Reuben Dorey

BLANDFORD PRESS
POOLE DORSET

First published 1976

Link House, West Street
Poole, Dorset BH15 1LL

ISBN 0 7137 0773 9

Printed in Great Britain by
Staples Printing Group Ltd
Colour plates printed by Tindal Press Ltd., Chelmsford

Contents

Acknowledgements vi
Foreword vii
Introduction ix
1 From Green Fingers to Test-tubes 1
2 Some Essential Facts of Science 4
3 The Plant Manufacturing Process 25
4 The Growing Medium and Environment 32
5 Starting the Plant 56
6 A Balanced Plant 60
7 Water and Chemical Supply 68
8 The Critical Stage 79
9 Full Production 89
10 The Fifth Column 98
11 Crop Programmes for North European Latitudes 106
Index 110

Acknowledgements

The author and publishers gratefully acknowledge the following sources of illustrations used in the book:

Imperial Chemical Industries Ltd. for colour plates 6–11;
Craig Engineering Ltd. for colour plates 1, 2 and 5
The Guernsey Press for Fig. 15
Jayglass Ltd., Guernsey for Fig. 12
Volumatic Ltd. for Fig. 25

Foreword

Since Reuben Dorey's book *Tomato Growing by Prescription* was published in 1960, the techniques of tomato production have changed in a number of respects. Perhaps the most significant has been the general adoption of carbon dioxide enrichment as an essential part of a grower's programme for producing early crops. Such crops are now expected to commence production in March and ultimately give total yields of some 250–300 tonnes per hectare.

This level of production has not, however, been achieved by merely adding the single new factor of carbon dioxide enrichment to the methods which were used up to 1960. As a consequence, other factors, especially temperature, have had to be re-investigated and new programmes or 'blueprints' devised which reflect a considerable amount of recent scientific research on the environmental requirements of the tomato plant. In this new book Mr. Dorey faithfully takes account of these changes and offers the reader an up-to-date statement of the techniques of tomato growing.

Another drastic change, sadly outside the grower's control, has taken place during the intervening years. There have been massive increases in the cost of fuel for heating glasshouses. This has radically changed the economics of tomato production. Because there are strict limits to the savings which can be made on growing costs, especially in regard to glasshouse heating, it has become an economic necessity for the grower to increase the cash value of his crop. Whilst an upward trend in market prices has helped in this respect there is still the need for the grower to consider how he can increase the productivity of his plants.

This is where a book such as Mr. Dorey has written can be of immense help because it outlines clearly, in simple language, what are the essentials for high level tomato production. It exhorts the reader to digest a few simple scientific facts and then to relate them to the problems of growing. But it does more than this. By frequent reference to the points of similarity between an industrial chemical manufacturing process and the chemistry of the tomato seen as a biological manufacturing process Mr. Dorey hammers home his most important message. It is that a high degree of precision is necessary with both processes if the end products are to be satisfactory.

It has always been obvious to the keen observer of the horticultural scene that the successful grower is one who pays great attention to detail and strives for the highest possible level of precision in his growing operations. This book will aid those growers who are prepared to recognise that they can and should do better. For the student and aspiring grower of the future it should be compulsory reading.

A. Calvert
April 1976

Glasshouse Crops Research Institute,
Littlehampton, Sussex

Introduction

In 1960, when *Tomato Growing by Prescription* was published, commercial tomato growers were beginning to realise that the application of scientific principles to their comparatively rustic methods of cropping was going to be of considerable benefit. Numerous research and experimental stations throughout the world were publishing data about plant growth and behaviour and the foundations of a 'blue print' method of tomato production were being laid. Running in parallel with this advancement in technical knowledge there began to appear new glasshouse structures, heating and ventilation equipment and chemicals for plant nutrition and pest control. Noteworthy among these developments was the introduction of means to enrich the glasshouse atmosphere with carbon dioxide at an economic cost.

The more progressive among the professional growers quickly appreciated the value of the new approach to their industrial problems and the consequent economic potential. 'Programme' growing has been generally accepted by them and their crops have improved not only in weight but in quality. Some details in the programme are still undergoing modification as new information comes forward but the main principles are firmly established. This book is an attempt to adapt those principles to the simpler, less sophisticated sphere of the smallholder and amateur gardener.

So often these growers are showered with well-meaning practical advice which, by its profusion, variety and occasional contradictions leave the recipient in a more confused state than before.

You may provide a grower with the most up-to-date glasshouse and modern equipment but all this will not produce a successful crop of tomatoes unless he knows how to use it. On the other hand a grower who has a comprehensive grip of the essentials of tomato plant physiology and of the environmental conditions necessary for optimum growth will know how to use quite rudimentary equipment with surprisingly good results.

This book deals with basic principles, starting with a little 'science made easy' which is then applied to the growing of a tomato plant. Practical methods based on the fundamentals of programme growing are briefly described and much of a controversial character has been deliberately excluded in the interests of simplicity. The programmes in Chapter 11 summarise the recommendations made in earlier chapters.

Reuben Dorey
April 1976

Castel, Guernsey

1 From Green Fingers to Test-tubes

Commercial culture of tomatoes under glass developed slowly over a period of seventy or eighty years up to the early 1950's. The pioneers were little more than skilled gardeners, but by intelligent application of their observations they laid the foundations of what was to become one of the most important sections of the horticultural industry.

The tomato crop was an annual one and many seasons had to elapse before new methods and improvements in technique could become established. However, when scientists began to take an interest in the tomato, the rate of advancement in knowledge increased considerably. It was natural that early experimental work should have been directed to those practical difficulties which hindered the production of fruit. Foremost among these were the losses caused by pests and diseases.

One of the most significant steps was the introduction of soil sterilisation into the seasonal preparation programme. This and the use of sprays and fumigants for attacking the pests present on the aerial portions of the plants did much to improve crop yields. Attention then turned to the many environmental factors which influence tomato fruiting. These included the temperatures for propagating and growing; watering and manuring. It soon became necessary for scientists to investigate more closely the fundamental nature of tomato plant physiology so that more information could be obtained about how the plants grow and react to their environment.

As a result of this experimental work, and the application of the information gained to their growing methods by progressive growers, great changes have taken place in commercial glasshouse culture in the last twenty years. It is now a highly scientific process with skilled technicians taking the place of green-fingered sons of the soil, producing crops that would have been considered quite unattainable some thirty years ago.

Tomato yields of 300 tonnes per hectare are now within the capacity of many professional producers but statistics show that the average rate of production is little more than half this quantity. There are many reasons for this, but as most of them are within the control of the grower, it is surprising that they lag so far behind the pace-setters when all they have to do is to put into practice a programme of operations that have proved so successful.

The small grower may not be able to afford many of the more modern systems of environmental control used on the larger industrial units and must therefore use his head and hands to carry out some of the necessary jobs. Yet if he knows exactly what conditions must be maintained in order to produce a high yield from his glasshouse, then confidence that his efforts will meet with success will be that much greater.

Equally, the keen amateur gardener with his small greenhouse will find more satisfaction by following the scientific methods which have been practically effective. Cultural methods based on knowledge of the growth processes in plants must apply to all growers, whether in the back garden or in the multi-hectare commercial holding. It must be an advantage for either type of producer to know as much as possible about how tomato plants grow in order to understand and put into practice the techniques most efficiently.

It may seem absurd to compare tomato production with a manufacturing process, such as, for instance, artificial silk synthesis. Nevertheless, this is no fanciful comparison as a brief examination will show.

The manufacture of nylon—a completely synthetic material—requires as raw materials the same substances as a tomato plant, namely the chemicals carbon, hydrogen, oxygen and nitrogen. A great deal of painstaking research in laboratories led eventually to the synthesis of the fibre we call nylon. When all the data concerning temperatures, gas and liquid pressures and chemical quantities had been collected in the laboratories a small-scale 'plant' was built, in which the processes worked out in glass apparatus were tried out in factory type equipment. At this stage more useful information was obtained about the factors which become apparent because of the increase in scale of the operation. Having done all this the manufacturer could then build his factory with the reasonable certainty that it would turn out the product nylon in predicted quantities and of the prescribed quality. Furthermore anyone in possession of the same information —and of course the necessary finance—could make exactly the same material.

Compare this method of manufacture with that of the tomato grower. He starts with the same raw materials as the nylon manufacturer, the same scientific laws and the same influences of temperature and other physical factors. But unless he happens to live in a climate where all these conditions are naturally favourable he will fail to produce tomatoes unless he takes a number of steps to control the environment. So he must study every stage in the process of building up the tomato plant from its raw materials and when he has found out how they work and what physical conditions are most favourable, then he can go ahead with confidence that having secured these conditions, a good tomato crop will be produced. Research is going on all over the world and the secrets of tomato plant physiology are yielding to the patient probing of scientists. Gaps in our knowledge are being filled and programmes can now be formulated which give growers firm guide lines along which they may proceed towards the production of their crops. If the grower is to understand fully these programmes, he will benefit from an acquaintance with some elementary facts about the science that lies behind them.

We have already learnt that the principal raw materials required for synthesising a tomato plant are the chemicals carbon, hydrogen, oxygen and nitrogen. What are these substances? Where do they come from and how do we manipulate them? In the next chapter we shall try to answer these and other questions as simply as possible.

2 Some Essential Facts of Science

It is almost impossible to discuss any aspect of plant culture without using scientific terms. Gardeners talk of nitrate, phosphate and potash and the average commercial grower may argue enthusiastically about the merits of osmotic control of plant growth or the effects of high pH. These are signs of the times; the scientific era is gradually superseding the green-fingered age.

For those who have neither the time nor inclination to study physics and chemistry in detail but who wish to acquire some knowledge of these subjects, an attempt will be made to give an outline of those facts that are helpful in an understanding of present-day growing methods. In a highly condensed version such as this must be, generalisations have to be made which, though not untrue, may depart slightly from the strictly accurate. However, as the object is to learn more of the intricate processes of tomato growth, this aim may justify the method adopted in learning something of the scientific background.

Chemistry

We begin with that branch of science which deals with the composition of matter.

When a grower thinks of chemicals he usually has in mind those bags of fertilisers that he buys for applying to the soil. However, tell him that the water he uses and the irrigation equipment through which it passes are also chemicals and he is liable to be somewhat sceptical. The fact remains that everything we see and touch is a chemical or mixture of chemicals masquerading under familiar names. By methods of chemical analysis all these common everyday materials can be broken down into simple substances known as chemical *elements*. Between ninety and a hundred separate chemical elements are known, about half of which occur in minute quantities. We need only concern ourselves with a very few of the more widespread substances. They are carbon, oxygen, hydrogen, nitrogen, sodium, potassium, calcium, magnesium, phosphorus, iron, aluminium, manganese, copper, zinc, sulphur and silicon. All these elements are found in the earth's crust or in the atmosphere and as constituents of plant and animal life. Some of their names will be quite familiar; for instance the metals iron, copper, zinc and aluminium. Carbon too, under the guise of coal or soot, is well known, though in its other form, the diamond, it may not be so easily recognised.

Oxygen, hydrogen and nitrogen are gases, available commercially in large quantities compressed in cylinders. The air we breathe consists mainly of a mixture of 4 parts of nitrogen and 1 of oxygen. Not so well known perhaps are the metals calcium, magnesium and manganese, while sodium, potassium and phosphorus are rarely seen outside chemical factories and laboratories. They are soft, cheese-like materials which have an unpleasant tendency to catch fire if left lying about. Sulphur was at one time used by growers for fumigating their glasshouses at the end of the season. Finally silicon, although rarely seen in its elemental form, is one of the most commonly occurring chemicals, chiefly in the form of quartz and sand.

The presence of these chemical elements as constituents of plants may be difficult to accept and it must be pointed out that they do so in combination with one another in the form of *compounds*. A chemical compound is a substance produced by the union of two or more elements under certain specified conditions. For example, a mixture of the gases hydrogen and oxygen could remain peacefully together at ordinary temperatures, but if subjected to a high temperature will combine with explosive violence to produce water. Carbon, as coal, may be seen in any coal fire combining with oxygen to form the compound carbon dioxide. This reaction is called combustion or burning and we meet it in plant growth but then the reaction is known as *respiration*.

Table 1 Chemical Symbols of Elements with Atomic Weights

Element	Symbol	Atomic Weight
Aluminium	Al	27
Boron	B	10
Calcium	Ca	40
Carbon	C	12
Copper	Cu	63
Hydrogen	H	1
Iron	Fe	55
Magnesium	Mg	24
Manganese	Mn	54
Molybdenum	Mo	95
Nitrogen	N	14
Oxygen	O	16
Phosphorus	P	31
Potassium	K	39
Silicon	Si	28
Sodium	Na	23
Sulphur	S	32
Zinc	Zn	65

The number of chemical compounds that it is possible to prepare from a selection of elements varies greatly and the scientific laws which govern their combination are very well defined.

The smallest particle of an element capable of existing by itself is called an *atom*. For a long time atoms were thought to be indivisible but we now know that they are composed of particles of positive and negative electricity and can be best understood as being arranged like a miniature solar system. The tremendous amount of energy released when atoms are broken up, as in nuclear bombs, gives some idea of the electrical forces which are locked up in these minute particles. We shall meet these forces again when we come to the subject of solutions. It is convenient however to regard atoms as indivisible since a great deal of useful information can be gained thereby.

Chemists have evolved a kind of shorthand in which each chemical element is denoted by a symbol (Table 1). An important point to note about these symbols is they also represent the weight of the atoms relative to one another. The difference in weight between lumps of iron and aluminium is obvious, but if we could weigh one atom of iron we would find that it was very nearly twice as heavy as one atom of aluminium. The atomic weights shown in the table are given in round figures for simplicity. In fact they vary from these figures by only small fractions, which can be neglected when using the numbers for practical purposes. It should be emphasised that they are not the actual weights of atoms but only relative weights. Thus if we express weights in grams then the symbol N would represent 14 grams of nitrogen; O would represent 16 grams of oxygen, and so on.

Atoms combine together to form compounds and the smallest particle of a compound capable of independent existence is called a *molecule*. For example, water is a compound of 2 atoms of hydrogen and 1 of oxygen, written with the familiar *molecular formula* H_2O. We can express this chemical reaction in the form of an equation:

$$2H + O = H_2O$$

From this equation we can derive the information that 2 parts of hydrogen (grams, kilograms, etc.) combine with 16 parts by weight of oxygen to give 18 parts by weight of water.

Another compound which is of great importance in plant growth is carbon dioxide. This gas is formed during combustion or respiration by the combination of 1 atom of carbon with 2 atoms of oxygen:

$$C + 2O = CO_2$$

Referring to the atomic weights table we learn that 12 g of carbon and 32 g of oxygen produce 44 g of carbon dioxide. It will be seen that, just as there are atomic weights for atoms so there are molecular weights for compounds, i.e. 18 for water and 44 for carbon dioxide.

It is now possible to calculate the amounts of chemicals taking part in any chemical reaction. For example, take the well-known plant food potassium nitrate; chemical analysis shows that it consists of 1 atom of potassium (K), 1 atom of nitrogen (N) and 3 of oxygen (O) with the molecular formula KNO_3. Its molecular weight is:

K	39
N	14
3O	48
	101

101 parts of potassium nitrate therefore contain 39 parts of potassium or 38·6% K. The nitrogen content is $\frac{14}{101} \times 100 =$ 13·8% N.

At this point it is worthwhile digressing to refer to the terms used in horticulture when evaluating the nutrient content of fertilisers. It has long been the custom to express the potassium content in terms of potash. Now potash is a name commonly given to the oxide of potassium K_2O, which is not what plant roots absorb. The same comment applies to the phosphorus content of manures which has always been expressed as 'phosphate' and written P_2O_5, the molecular formula of the compound phosphorus pentoxide. There are many phosphates and those used for plant nutrition are mixtures of various kinds. The use of potash (K_2O) and 'phosphate' (P_2O_5) to denote the potassium and phosphorus contents of fertilisers is so firmly and universally established that it might be confusing to make a sudden change and use the symbols K and P with their associated atomic weights. We shall therefore continue to use this terminology always bearing in mind that plants are only able to absorb potassium K and phosphorus as H_2PO_4. The inconsistency of horticultural chemists of the past and present is demonstrated by the fact that nitrogen is always referred to and calculated as atomic N whereas plant roots absorb nitrogen in the form of nitrate NO_3.

Studies of the mineral nutrition of plants have shown that of the eighteen elements listed in Table 1 the fifteen in heavier type are known to be essential for growth. The chemical reactions and changes which take place during the development of a plant require the presence of these fifteen elements in large or small amounts but some of them are merely helpers or catalysts in the complex reactions and do not form part of the structure of the plant.

Solutions

If a few crystals of a substance like potassium nitrate or common salt are dropped into a glass of water they soon disappear and we say they have dissolved and have formed a *solution*. What has happened to the solid?

It has been pointed out that atoms are electrical in nature, consisting of negative electrical particles, called *electrons*, surrounding a positive nucleus. In compound molecules the various electrical forces neutralise one another and their mutual attraction keeps the molecule rigid. Water has a weakening effect on these electrical bonds causing atoms or groups of atoms to separate into freely moving particles. Let us return to our crystal of potassium nitrate in the glass of water. The molecular formula of potassium nitrate is KNO_3 and in water the electrical bond between the potassium atom and the remainder of the molecule NO_3 is loosened. On separation there is a shift in the electrical balance; the potassium atom loses an electron (negative charge) and consequently becomes positively charged. This electron attaches itself to the NO_3 part of the molecule giving it a negative charge. These charged atoms or groups of atoms in solution are known as *ions* and the process is known as *electrical dissociation* or *ionisation*. Positive ions are called *cations* and negative ions are called *anions*.

Making use of our chemical symbols we can express the process of dissociation as follows:

$$KNO_3 \xrightarrow{\text{in water}} K^+ + NO_3^-$$

A solution of potassium nitrate and many similar compounds contains millions of cations and anions together with a few undissociated molecules of the original compound. Other compounds dissociate to a very limited degree when dissolved; in their case the molecules remain intact but the electrical forces which bind the molecules together in the solid state are weakened by the water so that the molecules are free to move about in the water. Just as the particles of a gas fill up the space in which the gas is enclosed, so the particles of a substance dissolved in water diffuse throughout the liquid volume until distribution is uniform.

A solution is characterised by the amount of substance dissolved in the liquid; this very important feature is known as the *concentration* of the dissolved substance, and may be expressed in several ways. In scientific work the usual terms are grams weight per litre volume (g/l):

$$
\begin{aligned}
1\ \text{g/l} &= 1\ \text{g in 1000 millitres (ml) water} \\
&= 0{\cdot}1\ \text{g in 100 ml water} \\
&\qquad \text{concentration } 0{\cdot}1\%
\end{aligned}
$$

When dealing with very weak or *dilute* solutions such as soil or plant nutrient solutions another method of expressing concentration is common, namely 'parts per million'. This is essentially a weight/weight relationship and is applicable to any system of measurement. A solution made up to contain 100 g of solid in 100 litres of water will have concentration of 1 g/100 g water. One million grams of this solution will therefore contain 1000 g of solid and we say that the solution has a concentration of 1000 parts per million, abbreviated to 1000 ppm.

In the feeding of tomato plants with liquid nutrient solutions the term *dilution* is constantly employed. For instance, a mixture of chemicals may be dissolved in water at the rate of, say, 200 g per litre. This solution is then applied to the plants through a diluting apparatus wherein water is added to produce a final solution having a much weaker concentration. Suppose the rate of dilution is 1 in 100; this means that 100 litres of the final solution will contain 1 litre of the original stock solution. The concentration of the diluted solution is found as follows:

1 litre stock solution contains 200 g solid

therefore 100 litres dilute solution contains 200 g solid

$$
\begin{aligned}
\text{i.e. concentration} &= 2\ \text{g/litre} \\
&= 2\ \text{g/1000 g} \\
&= 0{\cdot}2\%\ \text{wt/wt} \\
&= 2000\ \text{ppm}
\end{aligned}
$$

Acids, Bases and Salts

There must be few growers these days who have not at some time come across the substances sulphuric acid or nitric acid. These are two very familiar examples of *acids*, a group of chemicals which all have one thing in common; when dissolved in water they produce hydrogen ions (H^+). The remainder of the acid molecule consists of a group of atoms which is characteristic of the particular acid.

As examples we have:

$$\underset{\text{sulphuric acid}}{H_2SO_4} \xrightarrow[\text{dissociation}]{\text{on}} \underset{\text{hydrogen ions}}{2H^+} + \underset{\text{sulphate ion}}{SO_4^{--}}$$

$$\underset{\text{nitric acid}}{HNO_3} \longrightarrow H^+ + \underset{\text{nitrate ion}}{NO_3^-}$$

$$\underset{\text{phosphoric acid}}{H_3PO_4} \longrightarrow 3H^+ + \underset{\text{phosphate ion}}{PO_4^{---}}$$

Another group of chemicals of importance includes the substances lime (calcium hydroxide $Ca(OH)_2$) and magnesia (magnesium hydroxide $Mg(OH_2)$. These are known as *bases* or *alkalis* and in water all produce the anion called hydroxyl (OH^-)

$$\underset{\text{potassium hydroxide}}{KOH} \xrightarrow[\text{dissociation}]{\text{on}} \underset{\text{potassium ion}}{K^+} + \underset{\text{hydroxyl ion}}{OH^-}$$

$$\underset{\text{calcium hydroxide}}{Ca(OH)_2} \longrightarrow \underset{\text{calcium ion}}{Ca^{++}} + \underset{\text{hydroxyl ions}}{2OH^-}$$

Acids and alkalis have a great affinity for one another and pair off on the slightest pretext. The products of these unions are called *salts*; some common examples are the salts potassium nitrate, potassium sulphate, calcium phosphate, magnesium sulphate and ammonium nitrate, all used in plant nutrition.

Let us write down the equation showing the chemical reaction between an acid and a base in the production of a salt:

$$HNO_3 + KOH = KNO_3 + H_2O$$

nitric acid — potassium hydroxide — potassium nitrate — water

In solution, where the substances would be ionised, the equation would look like this;

$$K^+ + OH^- + H^+ + NO_3{}^- = K^+ + NO_3{}^- + HOH$$

It will be seen that the ions on both sides of the reaction equation are the same with the exception of the hydrogen and hydroxyl ions which combine to form undissociated HOH or water H_2O. If we wish to isolate the salt in solid form all we have to do is to remove the water by evaporation. The remaining ions then become attracted to each other again and unite to form crystals of solid salt.

Water

In more ways than one water is a mystery of nature. It is vitally necessary for the support of plant and animal life; it is capable of dissolving a very large number of chemicals but is itself almost completely undissociated; that is to say, the ionisation reaction:

$$H_2O \text{ or } HOH \rightleftharpoons H^+ + OH^-$$

proceeds to the right to a negligible extent and as a result the number of free hydrogen and hydroxyl ions in pure water is extremely small. In more precise terms, the concentration of H^+ ions is only one ten-millionth of a gram equivalent per litre of pure water. Those familiar with mathematical notation will know that this quantity may be written 10^{-7}. This figure has considerable significance for the grower.

pH

Acidity in a solution is produced by the presence of hydrogen ions. In pure water, H^+ ions are balanced by an equal number of OH^- ions and because of this the water is neither acid nor alkaline, but neutral. If we add to pure water a substance that will produce H^+ ions, then their concentration will increase and the solution will become acid. Measurements may show an increase in H^+ ion concentration to 10^{-6} or even 10^{-5} gram equivalents per litre. Because these concentrations are so low and the method of expressing them is clumsy a scale has been devised known as the pH scale which is much more convenient. Once again the mathematicians will know that the logarithm of 10^{-7} is -7 and it follows that the negative logarithm of 10^{-7} is 7. The negative logarithm of the hydrogen ion concentration is called pH, and at neutrality when hydrogen and hydroxyl ions are present in equal amounts:

$$pH = -\log(H^+) = 7$$

Table 2 shows the progression from an acid to an alkaline solution in terms of pH.

Table 2 pH Scale

Strongly acid	H^+ ion concentration 10^{-4}	pH 4
Very acid	H^+ ion concentration 10^{-5}	pH 5
Acid	H^+ ion concentration 10^{-6}	pH 6
Neutral	H^+ ion concentration 10^{-7}	pH 7
Alkaline	H^+ ion concentration 10^{-8}	pH 8
Very alkaline	H^+ ion concentration 10^{-9}	pH 9

The alkalinity of a solution with a pH above 7 is due to the presence of an excess of OH^- ions.

Electrical Conductivity

In a salt solution, positive and negative ions are moving about freely in the water. If we place two metal plates in such a solution and connect one to the positive terminal of a battery and the other to the negative terminal the cations (+) will be attracted to the negative plate and the anions (—) to the positive plate. By doing so an electric current is transmitted through the liquid. The more ions there are in the solution the greater the current; or, the *electrical conductivity* of the solution varies according to the concentration of salt ions. By measuring the electrical conductivity of a solution we have a means of assessing the strength or concentration of the solution. This measurement has now become an important feature in soil analysis.

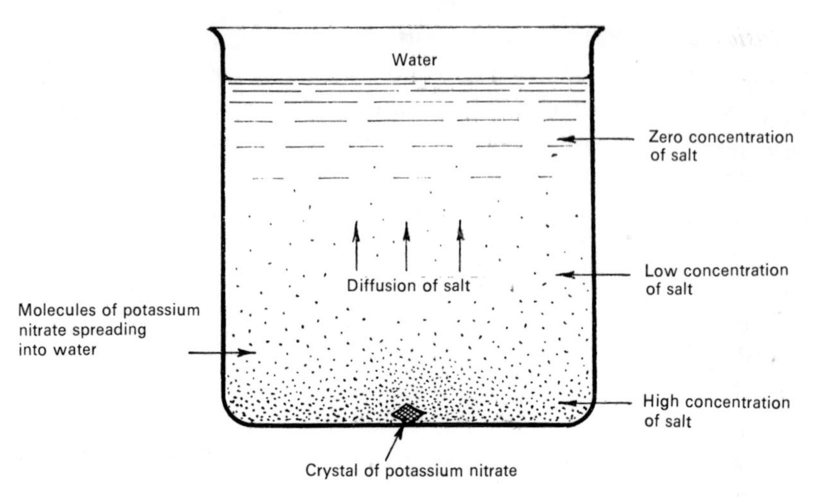

Fig. 1 Diffusion of a dissolved salt from a region of high concentration to one of low concentration.

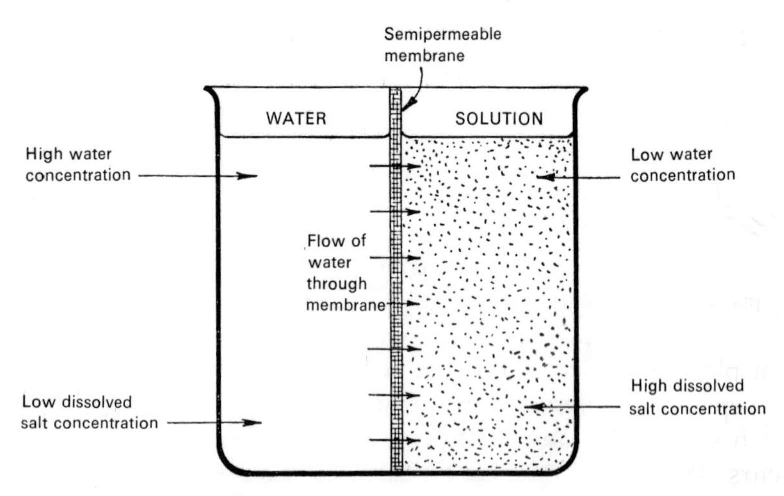

Fig. 2 Osmosis—Diffusion through a membrane. Water flows from high to low concentration. The dissolved salts which would diffuse from right to left down the concentration gradient are prevented from doing so by the membrane.

Diffusion and Osmosis

Water always flows from a high level to a lower when left to itself. Similarly a gas will fill a space under the influence of pressure. Imagine a sphere completely empty, that is, without air. If by opening a small valve for a fraction of a second the interior of the sphere is put in contact with the air outside a small volume of air will pass into the container. It will do so because the atmospheric pressure is higher than that inside the sphere. But the air that passes inside will not stay in one little 'lump' at the bottom; it will gradually fill up the whole volume until the pressure is uniform throughout.

Particles dissolved in a liquid behave very much like gas particles and will move from a region of high concentration to one with a low concentration. This movement of gas or dissolved particles from one place to another is called *diffusion* and it will continue as long as there is a difference in concentration between one part of the space and another.

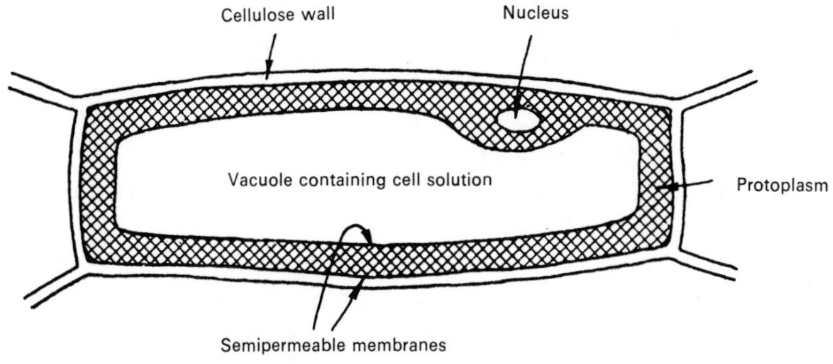

Fig. 3 Diagram of typical plant cell.

In plant physiology there are many instances where solutions are separated from one another by thin membranes through which water and some dissolved substances may pass. Diffusion occurs at a reduced rate and the permeability of the membranes varies for different substances. The name *osmosis* is given to the diffusion of water through semi-permeable membranes.

We find these membranes in plant cells, which consist of a

layer of protoplasm surrounding a cavity of vacuole, the whole being enclosed in a fairly rigid cell wall composed of cellulosic material. On each side of the protoplasmic layer are two thin membranes which are the principal barriers to the passage of dissolved substances. Water passes through these membranes easily and quickly. The cell vacuole contains a liquid, or sap, which is a solution of sugars, proteins and inorganic salts. Cells are joined together by carbohydrate substances permeable to water and dissolved chemicals; thus it is possible for water and chemicals to diffuse from one cell to the next in accordance with the concentration gradients in adjacent cells.

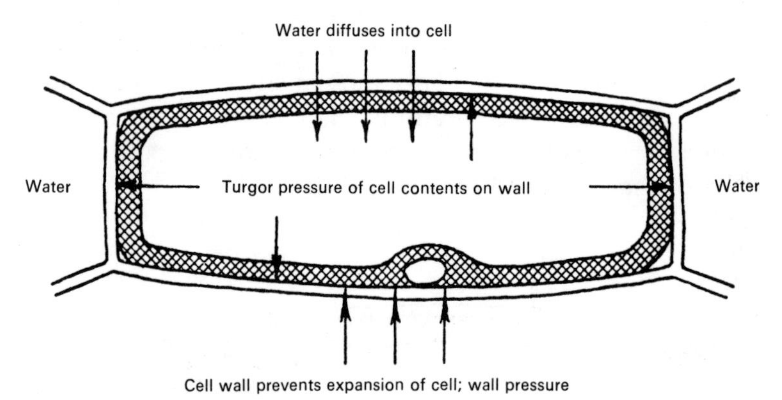

Fig. 4 In a turgid cell, the osmotic pressure of the cell contents is balanced by the cell wall pressure.

If one of these cells is placed in pure water there will be a difference in concentration between the solution in the cell and the water outside. Water will pass from the region of high water concentration outside the cell into the cell where the water concentration is low. This will cause the cell to swell and a point will be reached when the cell wall presses back on the cell contents with a force that prevents further entry of water. This pressure is the *osmotic* pressure of the cell solution.

Suppose now that such a cell is placed in a solution having a salt concentration higher than that of the cell sap. The water concentration gradient is now reversed and water will flow out of the cell. As a consequence of this outflow of water the cell will lose its turgidity and the inner membranes may shrink away from the cell wall. Such a cell is said to be *plasmolysed* and, unless the process has gone too far, it can be restored to its turgid state by diluting the external solution.

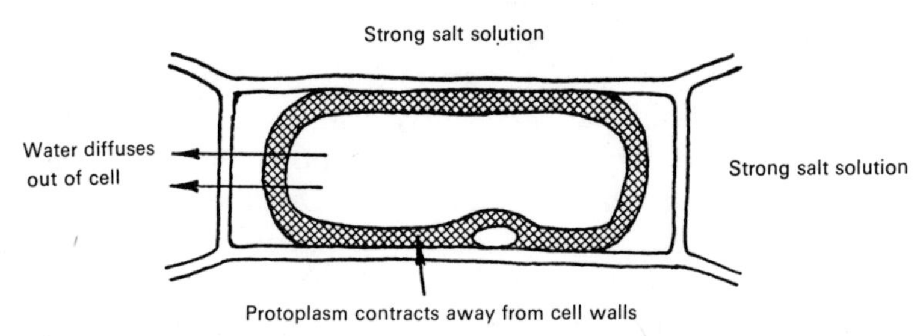

Fig. 5 Plasmolysis of a cell placed in a strong salt solution. A cell left in this condition will die.

Osmosis, or diffusion through cell membranes, is responsible for passage of water from cell to cell in a plant. The absorption of ionic chemicals by roots from the soil solution is a somewhat different process. In this case ions enter the root cells against a concentration gradient but in order to do so a certain amount of energy must be used. You cannot put more air into a bicycle tyre without doing some work with a pump. Active salt absorption can only take place if root cells are provided with a plentiful supply of oxygen and carbohydrates for vigorous respiration and the generation of energy.

Carbon Compounds

The compounds of carbon have a special place in chemistry. There are hundreds of them and because so many are found in the structure of plants and animals they have been given the name *organic* chemicals.

Carbon has an almost insatiable affinity for other chemical elements and for hydrogen, oxygen and nitrogen in particular. It would be quite impossible to do more than refer to some of the more important groups of compounds which occur in plant growth.

Of these the *carbohydrates* are probably of major interest. They are compounds of carbon, hydrogen and oxygen and include the sugars, starches and celluloses. The simple hexose sugars have the general molecular formula $C_6H_{12}O_6$ and a well-known example is glucose. The sugars are complex substances as may be gathered from the fact that they contain so many atoms. It is possible to combine these atoms in many different ways so that there are a number of different sugars with the same basic formula. The molecules of simple sugars may also unite to produce even more complicated larger compounds known as *polysaccharides*. Sucrose, the sugar which is much concerned in plant growth, contains two hexose molecules and therefore has the formula $(C_6H_{12}O_6)_2$. The celluloses which form part of plant tissues are very large chain-like aggregations of sugar molecules.

The second important group of organic compounds which play a leading role in plant life consists of the *proteins*, characterised by the presence of nitrogen in their constitution. Free nitrogen in the air is converted by bacteria in the soil into protein compounds, which in turn are transformed by other bacteria into simple nitrates.

A third class of carbon compounds comprises the fats and oils. In addition to forming reserves of food material for the plant, they are sometimes of considerable value to animals and man. Familiar examples of the latter are olive oil, cottonseed, linseed and castor oils.

Energy

Having taken a cursory look at the constitution and behaviour of some of the materials which the grower has to handle, it is necessary to pay some attention to that branch of science, called physics, which deals with the manner in which matter reacts to its environment. Water relations in plant growth, the influence of temperature and light and the practical techniques of glasshouse heating cannot be fully understood without some knowledge of the basic principles involved.

When we come to discuss the subjects of photosynthesis and

respiration we shall find ourselves constantly using the term 'energy'. The usually accepted meaning of the word, indicating vigour or force, is not precise enough for the scientist. He recognises that there are a number of different forms of energy and has devised methods to measure them. His studies have led him to propound a law which appears to be one of the most fundamental in nature. It is called the *Law of Conservation of Energy* and states that energy is indestructible. In effect this means that one form of energy may be converted into others without loss.

A simple experiment may illustrate the scientific meaning of energy better than elaborate definitions. Imagine a loose brick in the parapet of a high building; if given a push the brick will fall to the ground. While it is still in position on the parapet the brick is said to possess *potential energy* by virtue of its elevated position above the ground. This energy is measured by multiplying the weight of the brick and its height above the ground. As soon as the brick begins to fall under the influence of gravity this potential energy is converted into energy of motion or *kinetic energy*. This kinetic energy will increase quickly as the brick gains speed until the moment of impact with the ground. The accumulated energy will then be dissipated into some other form of energy depending on the manner in which the brick completed its fall. For instance, it could be converted into useful work by driving a post into the ground. If it merely hit the ground and came to rest then it would

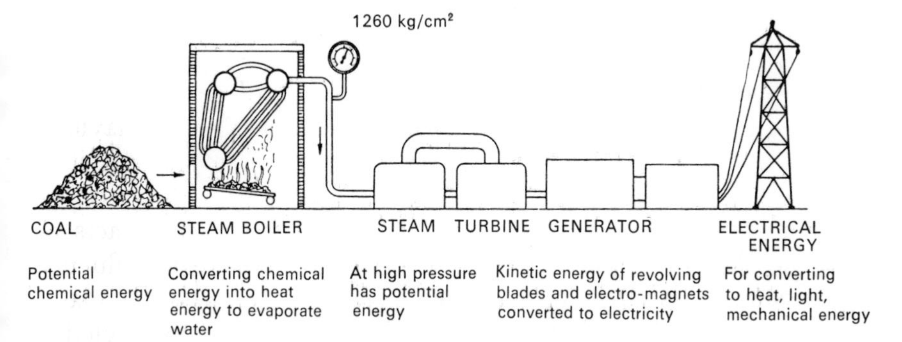

Fig. 6 Diagrammatic representation of energy conversions. The electrical energy obtained can be calculated from the amount of coal burnt, taking into account the efficiency of the different stages.

be found that the temperature of the brick and the surface of the ground had risen slightly. In other words kinetic energy can be made to do work and produce *heat energy*.

A coiled spring possesses potential energy because in returning to its uncoiled state it can drive the mechanism of a clock. Steam under pressure has potential energy because in expanding it can operate a piston which turns a wheel. The turning wheel represents kinetic energy which can be converted into *electrical energy* when the wheel is coupled to a dynamo.

Heat energy is one of the most important forms as far as the grower is concerned, and therefore deserves a little closer attention.

We have seen in the section on chemistry that matter is composed of compound molecules. In the solid state these molecules are held by electrical forces in fixed positions. Even so they are in constant vibration about these points. The heat content of a substance is associated with these molecular vibrations. If the speed of vibration increases a rise in temperature will be apparent. As the substance becomes hotter and the vibrations become more violent a stage will be reached when the molecules will slip away from their fixed centres of vibration and will move past each other. We say that the substance has melted (or fused) and become liquid. Further application of heat energy to the liquid leads to a greater agitation of the molecules and some escape from the surface of the liquid in the form of vapour. Eventually the whole of the liquid may be transformed to the vapour state. The changes from solid ice through water to steam are familiar to everyone. They are the outward signs of changes in the kinetic energy of water molecules brought about by the absorption of heat energy.

Evaporation

The escape of fast-moving molecules from the surface of a liquid goes on continuously and is known as *evaporation*. We have seen that the temperature of a substance is associated with the speed at which the molecules vibrate. When the fast-moving molecules escape from the surface the average speed of the remainder is

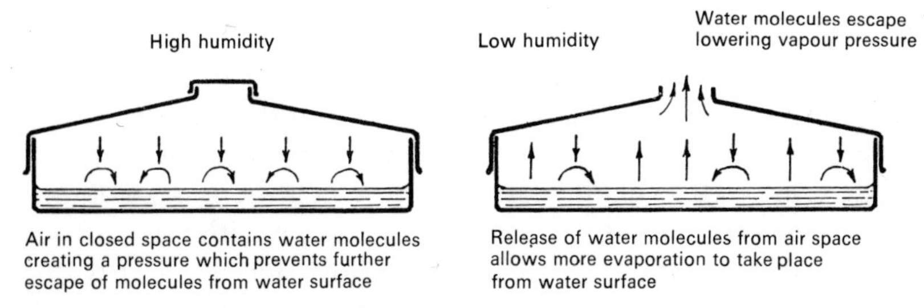

Fig. 7 Evaporation into enclosed spaces.

lowered and the liquid becomes cooler. Blowing across the surface of a hot liquid increases the rate of evaporation by carrying away molecules of liquid and preventing their return to the surface. The cooling effect produced on the skin by blowing is an example of the increase in evaporation rate.

Humidity

In addition to the gases oxygen and nitrogen, air also contains water vapour produced by evaporation of water on the earth's surface. The amount of water vapour in the air depends on the temperature. Every grower has seen evidence of water vapour in his glasshouses. In the early morning after a cool night the fruit and plants will be damp. During the night the air in the enclosed house will have become saturated with vapour and as the temperature drops it is unable to hold any more. Condensation of excess water vapour then occurs on cool surfaces. As soon as the temperature is raised, either by the sun or by the heating system, the capacity of the air to hold water vapour is increased and evaporation begins again. The plants dry out and although the atmosphere appears to be drier, it does, in fact, contain more water vapour than before. The ratio of the amount of water vapour in the air to the amount required to saturate it at the same temperature is known as the *relative humidity*. Under normal conditions the air contains about half the amount of water vapour that would be required to saturate it, i.e. the relative humidity is about 50%. In glasshouses it is usually somewhat higher, rising to 90% at night when the ventilators are closed and the temperature falls.

Energy of Chemical Reaction

All chemical reactions are accompanied by changes in heat energy. Easily observed are the heat evolutions from the burning of coal or oil and the slaking of quicklime (CaO) to form calcium hydroxide ($Ca(OH)_2$). The rusting of iron, in which there is a combination between iron and oxygen, also takes place with the evolution of heat, but since the process is a very slow one the heat is dissipated without perceptibly raising the temperature of the metal. However, if finely divided iron filings are dropped into pure oxygen the reaction is so quick that the particles become white hot.

The most important reactions with which the grower is concerned are those in which carbon and oxygen combine together or split apart. We can write this symbolically:

$$C + O_2 \rightleftharpoons CO_2$$

The reaction towards the right is the process of combustion or respiration and produces heat energy. The reverse reaction occurs in plants and requires an input of energy which comes from the sun. This is the largest natural industrial process in the world and is called *photosynthesis*.

Light Energy

The role played by light in plant growth is of such significance that it demands some consideration of the scientific facts behind a phenomenon that is often taken for granted. What is light and how does it produce such far-reaching effects? A full answer to this question is much too complex to be attempted in this book and we can only hint at some of the essential features that are relevant to our particular interest.

If we heat a piece of metal, say a poker, we shall notice that after a short time the metal is hot enough for the heat to be felt at some distance from the metal. Heat energy in the metal is being radiated from the hot metal and is detected without contact being made. The heat would still be felt if the metal were in a vacuum indicating that it is not the air which conducts the heat. As its temperature rises the metal begins to glow red, then orange and finally becomes 'white' hot. We have produced light by raising

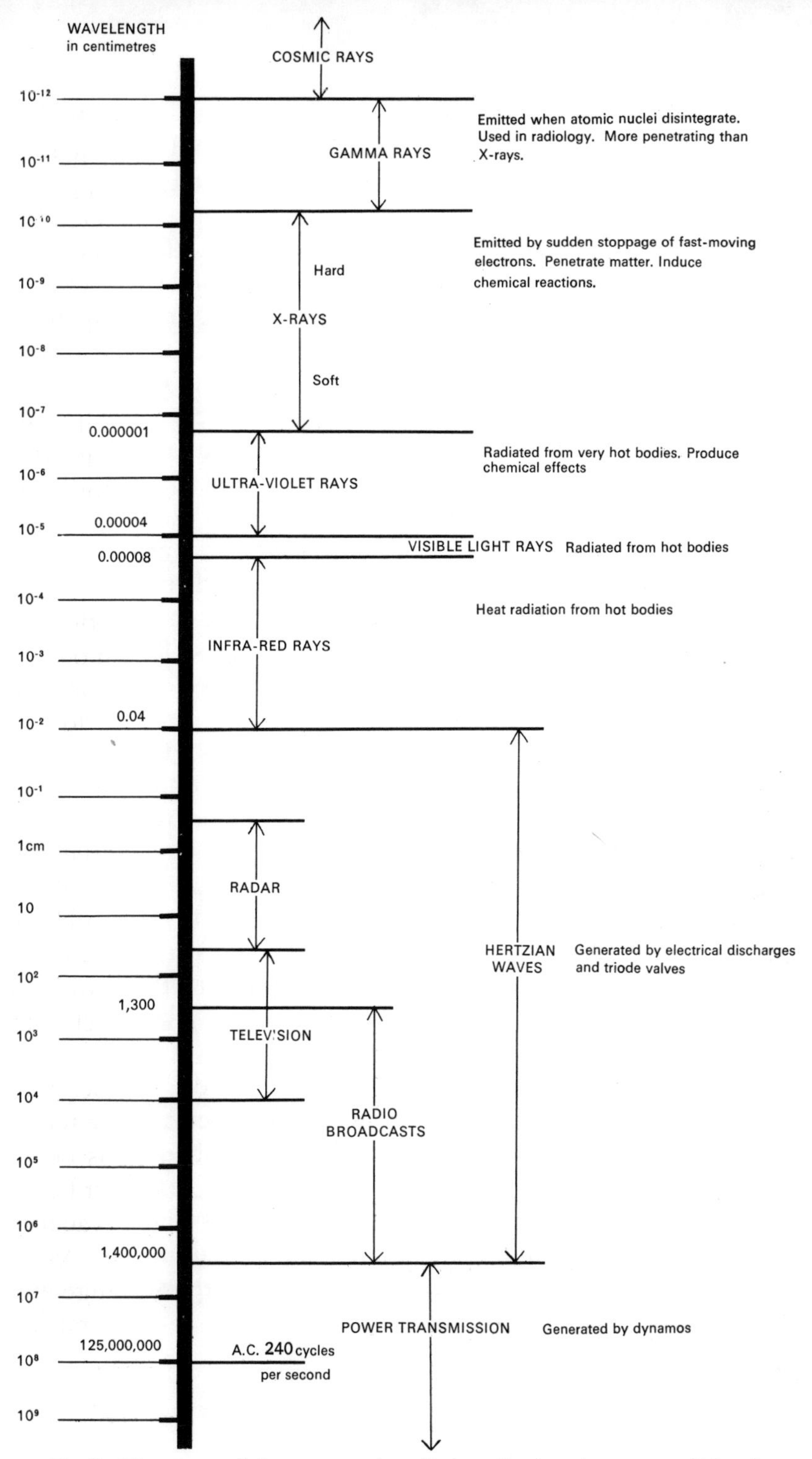

Fig. 8 The range of electromagnetic radiation showing the very small band which is visible light.

the temperature of the metal by several hundred degrees. The radiation that could be felt as heat before the metal began to glow is exactly similar in nature to that which produced the sensation of light.

The opposite reaction, namely the conversion of light energy into heat, can be illustrated when a glass lens is used to focus light rays on to a piece of paper. The radiation coming from the sun is concentrated on to one spot which becomes so hot that the paper burns.

Without attempting to give what would be a very involved and confusing explanation, it must be sufficient to state that the radiation we call light is but a small part of a universal form of electromagnetic radiation that embraces a vast range of physical phenomena. Radiant energy is propagated in waves having measurable wavelengths; it has its origin in deep-seated electrical changes in the structure of atoms which the reader will remember are composed of positive and negatively charged particles. The quality of radiation is described mainly in terms of its wavelength and the colours of the rainbow, or visible spectrum, are evidence of different wavelengths. The wavelength of red light is longer than that of blue light. As wavelength increases beyond that of red light the radiation becomes invisible but we can detect it by heat and infra-red photographic effects. The length of waves in the visible portion of the electromagnetic spectrum ranges from 0·00004 to 0·00008 cm. On either side of this very short band there is a continuous band of radiation extending, on the one hand, to wavelengths of 10^{-14} cm and on the other to the very long wavelengths of thousands of metres. Figure 8 shows the spectrum of electromagnetic radiation and indicates the uses to which the various bands are put in technology and in our daily lives.

The powerful atomic changes which produce this radiation give it a high energy content which when adsorbed by atoms of receptive substances can bring about chemical reactions. Such reactions are known as photochemical reactions. Photosynthesis, in which carbon compounds are manufactured in a plant, is stimulated by radiation from the visible part of the spectrum and in particular by the red and blue wavelengths.

3 The Plant Manufacturing Process

Our object is to study the natural growth of a plant as though it were a process of chemical synthesis under the full control of the grower. Whether he is a professional with maximum output per unit area of glass as the aim, or a gardener producing tomatoes for his family, precise information must result in efficient production and a predictable result. We must now look more closely at the chemical and physical processes which take place as the plant grows.

Plants grow as a result of a complicated series of chemical reactions which, under natural conditions, proceed without human intervention and which are subject to environmental control factors such as light, temperature, water availability, and atmospheric gas—all features of what we call climate. They will grow best, and sometimes only, when climatic conditions are precisely favourable. The tomato is a native of sunny, warm regions and grows best at temperatures between 14°C and 26°C. Commercial production in Northern Europe and the United States can only be conducted with greatest efficiency under protected conditions and thus requires heated glasshouses.

It has already been stated that the raw materials for producing a tomato plant are the chemicals carbon, hydrogen, oxygen, with potassium, phosphorus, calcium, magnesium and a few others to assist in the numerous chemical reactions. Fortunately for the grower an initial supply of some of these chemicals is already present in the tomato seed where the manufacturing process begins.

Nothing will happen to these chemicals in the seed until conditions are favourable for reaction to commence. This initial reaction is called *germination* and for it to be successful the environment surrounding the seed must be suitably moist, at the correct temperature and contain an adequate supply of oxygen.

The chemicals which are locked up in the seed in the form of fats, oils and proteins then begin to break down and the simpler chemicals that are formed react with oxygen from the air to produce heat energy. Further chemical changes follow with the

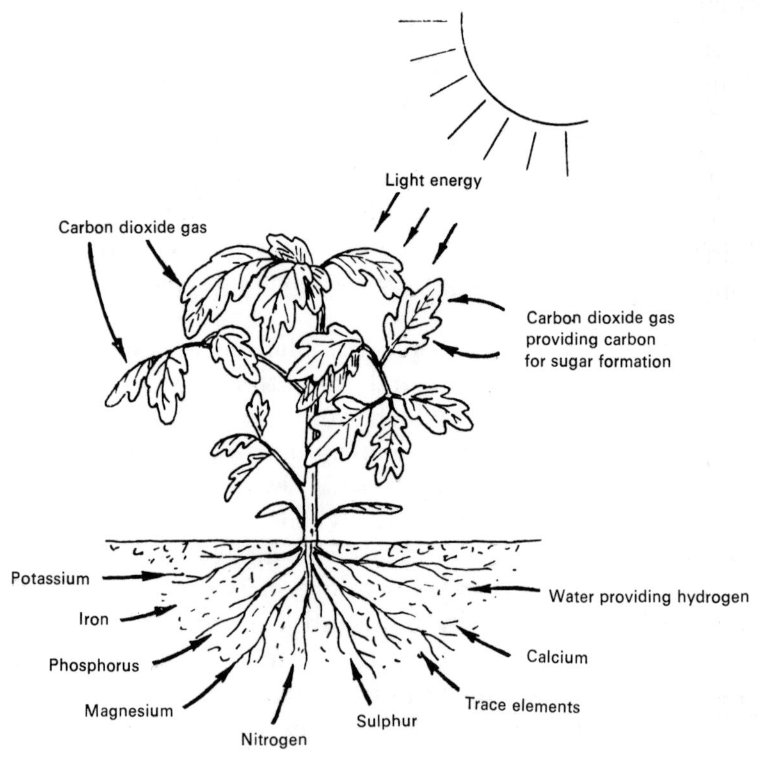

Fig. 9 Some of the raw materials required to build a plant.

formation of new plant material and the small seedling is on its way. Optimum conditions for germination are provided by a moist, open-textured medium maintained at a temperature of 18 to 21°C.

New plant tissue, the fabric of a plant, is formed by the division of microscopic cells in the growing points, of which there are two in the embryonic seedling. One growing point is destined to become a root and the other is the shoot or aerial portion of the plant. As the cells enlarge and divide the root and stem elongate. Very soon new cells are produced which become specialised and will eventually form leaves and part of the stem system used for transporting water and nutrients throughout the plant.

Photosynthesis

As soon as the shoot appears above the growing medium the first seedling leaves (cotyledons) begin to open out and spread themselves to catch the light.

The cotyledons are pale yellow at first but after a few hours in the light they turn green due to the formation of the green pigment *chlorophyll*. Chlorophyll is a large and very complex molecule containing about 135 atoms of carbon, hydrogen, nitrogen and 1 atom of magnesium. The elements iron and manganese also

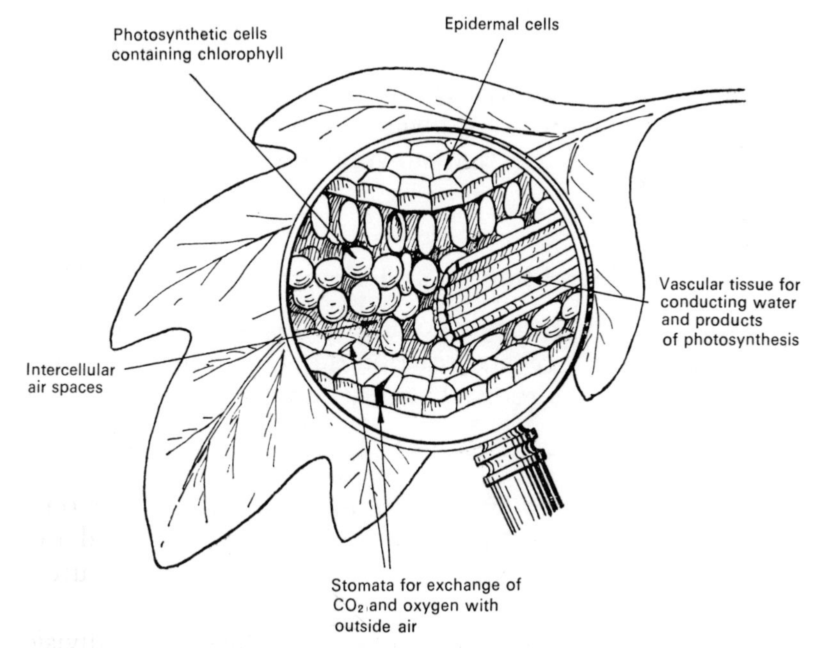

Fig. 10 Section through leaf showing structure adapted to the photosynthetic process.

take part in the formation of chlorophyll and deficiencies of the elements in the nutrition of the plant show up as yellowing of the leaves. Light energy is absorbed by chlorophyll and converted into heat which drives the entire process of photosynthesis.

The photosynthetic process involves the combination of carbon with hydrogen and oxygen to form carbohydrates and other organic substances. The carbon comes from the gas carbon dioxide (CO_2) present in the air; the hydrogen and oxygen come from water present in the leaves.

Air contains about 0·03% of CO_2 gas, some of which passes through minute pores in the leaves. This carbon dioxide dissolves in the water permeating the intercellular spaces of the leaves and the resulting solution finds its way into the photosynthetic cells. Here the energy produced by the action of light on chlorophyll causes a reaction to take place between CO_2 and water:

$$CO_2 + H_2O \xrightarrow[\text{energy}]{\text{light}} \underset{\text{sugar}}{(CH_2O)} + O_2$$

Three external factors affect the rate of photosynthesis: light, temperature and the concentration of CO_2 in the atmosphere. It is most important for the grower to ensure that none of these factors becomes limiting, thus slowing down the development of the plant.

Respiration, Mineral Supply and Transpiration

While carbon assimilation is going on in leaves and plant tissue is being built up, the reverse reaction is going on in other parts of the plant, notably the roots. This reaction:

$$(CH_2O)_n + O_2 \longrightarrow CO_2 + H_2O + \text{chemical energy}$$

is part of the process known as *respiration*. It goes on continuously in roots, stems, leaves, and even flowers and fruit, and is the source of energy required for a large number of chemical changes, including the vital process of cell division.

Photosynthesis and respiration proceed during the hours of daylight but at night photosynthesis stops. The rate of carbon assimilation can also be reduced appreciably if the plants are subjected to long periods of low light intensity. For continued growth to be maintained it is necessary that the amount of CO_2 converted during the day should exceed that used up by respiration during the twenty-four hours.

Approximately 48% of the carbohydrate manufactured in the leaves is transported down to the roots, where it is used to provide energy for the absorption of chemicals from the soil or rooting medium. As the respiratory process requires a considerable quantity of oxygen the presence of air around the roots is essential.

The synthesis of proteins, cellulose and a host of other chemicals rivals the synthesis of nylon but they are all performed by the plant with amazing ease. Hundreds of different chemicals are made, used and broken down during the growing cycle. Controlling and regulating the development of leaves, flowers and fruit are the substances known as *hormones*. These are special chemicals which are synthesised in various parts of the plant and used to integrate the activities of one area with those of another.

There is still a widespread belief among growers that the most important factor in growing a heavy crop of tomatoes is the fertiliser they apply to the soil. A look at the composition of plants soon puts mineral nutrition in its proper place. While different parts of a plant have varying analyses, the overall average dry weight is only about 10% of the fresh green weight; in other words, 90% of the total weight of a plant is water.

When the dry material is analysed we find that at least 90% of this is carbon, hydrogen and oxygen—chemicals derived from air and water. The remaining 10% of dry matter consists of nitrogen and the various mineral elements obtained from the soil. Thus, the chemicals from the soil (apart from water) total up to only 1% of the green weight of the plant.

The fact that this proportion is so small does not, of course, imply that it is unimportant. Mineral nutrition is an indispensable part of the manufacturing process, the chemicals playing vital roles in the elaborate mechanism of plant synthesis. These essential chemicals are potassium, nitrogen, phosphorus, calcium, magnesium, sulphur, iron, zinc, manganese, copper, boron and molybdenum. The first six are required in relatively large quantities, while the rest are only necessary in very small amounts and are sometimes known as *trace elements*.

Making all these chemical elements available to the plant roots in sufficient quantity is not as simple as it might appear to be, as we shall see later. A great deal depends on the type of rooting medium and the form in which the chemicals are present therein.

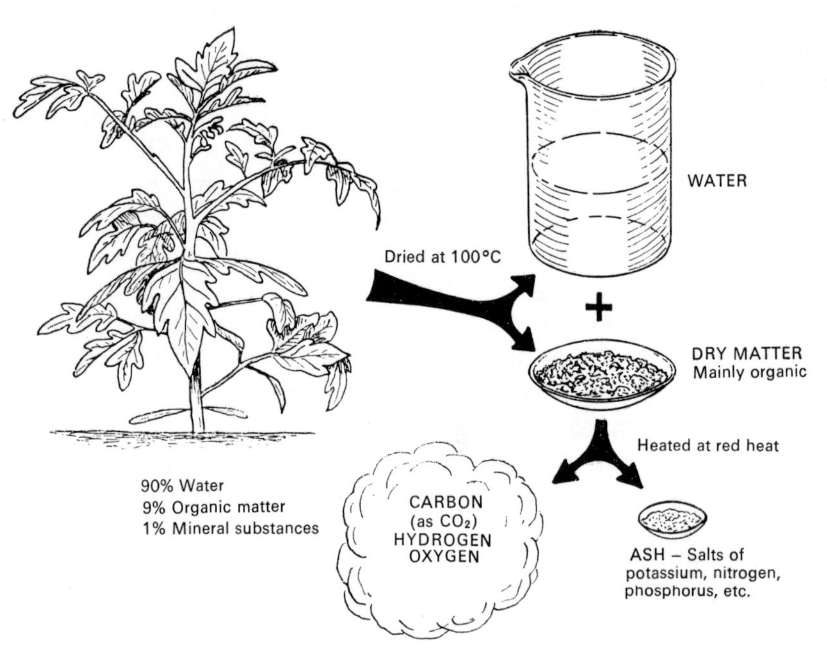

Fig. 11 The composition of a tomato plant.

Water is the medium in which all the chemical reactions occur within the plant and without if the whole manufacturing process would cease to function. While some of the water taken up by the roots is retained within cells and tissues the bulk passes through the plant and out into the air from the leaves. This flow of water is known as *transpiration* and it serves not only to translocate chemicals but to keep the plant cool by evaporation into the air. One of the most persistent tasks of the grower is to meet the demands of transpiration by irrigation.

We have outlined the essential processes in the manufacture of tomato plants and they are all within the capacity of the grower to control or influence in varying degrees. There are a few stages in the development of the plants when things seem to happen without apparent cause. For instance, the appearance of fruiting trusses in regular positions on the plant, the 'setting' of fruit and its subsequent swelling and ripening would seem to be stages which fall outside the grower's sphere of influence. But each of these 'natural' events is initiated by chemicals especially made for the purpose and because they are chemical reactions they are subject to the environmental conditions maintained by the grower. More will be said about these phases of development in later chapters.

4 The Growing Medium and Environment

A factory for making nylon or any other synthetic material often looks like nothing more than a conglomeration of large tanks and peculiar vessels connected together by a maze of pipes and tubes. Very little can be seen of the contents of this labyrinthine structure except at the few points where chemicals are introduced or extracted. Occasionally a trace of vapour may be seen—or smelt—but it is only at the end of the line that the final product makes its appearance. At various stages thermometers and control gauges are situated to allow an operator to check that internal conditions are maintained at prescribed values. Control valves are linked with these instruments and may be brought into action automatically thus reducing manpower and also the possibility of human error.

In the manufacture of tomato plants there is more to see and the grower has some advantage over his counterpart in the chemical factory. The plants themselves are the chemical reactors and all the grower has to do is to provide suitable means for the supply of water and nutrient chemicals and a controllable environment for the aerial portion of the plants.

The necessary features of these two media are as follows:

Atmospheric medium

(a) Protection from unfavourable weather conditions
(b) Maximum light
(c) Means to maintain temperatures within specified limits
(d) Apparatus to enrich the CO_2 content of the air
(e) Means to control humidity of the atmosphere

Rooting medium

(a) Adequate water supply and irrigation equipment
(b) Supply of nutrient chemicals
(c) Free flow of gases into and from the medium

In addition to the above essential requirements it is very necessary that both media should be free from pests and disease organisms. It is an unfortunate fact that the conditions which are most favourable for plant growth are those for the proliferation of these organisms.

Aerial Medium

Although tomatoes are grown in the open over a wide range of latitudes such crops are short-lived and light in yield. An average outdoor crop in Mediterranean countries and even in Jersey in the Channel Islands is no more than 1–2 kg per plant. On the other hand in properly constructed glasshouses and with careful control of all growth factors crops of 30 kg per square metre, or 10 kg per plant, are possible. Such crops are harvested from February to October from plants sown in November. The season is a year long and the change-over at the end of one crop to the start of the next must be very speedy.

Many growers and gardeners will not wish to emulate the performance of the highly sophisticated commercial enterprises and will be satisfied with a shorter and less demanding production season. Even so, they will need crop protection in the form of a glass or plastic house. The moment a structure is erected over a site the amount of light falling on that area is reduced. Measurements of light transmission through glasshouses show that there is a loss of 30–50% of the incident light. Structures should therefore be designed and erected in such a manner that transmission losses are minimal.

At one time glasshouses were based on a timber frame and although good wooden houses are still being built it is probable that the majority are now metal based. A whole volume could be written on the various types of structure available and it would

Fig. 12 A modern aluminium twin-span commercial glasshouse covering approximately 650 m^2.

only confuse the reader to mention more than the principal features to bear in mind when studying a builder's brochures.

Cost of construction is very high and must play a part in the choice but the cheapest is never likely to be the best. One of the first things to look for is the maximum area of glass in relation to the framework, and metal structures offer advantages in this respect. The side walls should be at least 2 m high to the eaves. The height to the ridge depends on the width of the house but where high summer temperatures are prevalent the pitch of the roof should not be too low. There is some evidence that some glasshouse blocks with narrow spans and low-pitched roofs tend to become uncomfortably hot in summer when days are long and sunny. During the past ten years or so glasshouses of many shapes and sizes have appeared. There was a 'wide-span' phase when huge houses with clear spans of 25–30 m were constructed. They were very good transmitters of light and it was possible to maintain extremely congenial air conditions during the summer. With the massive increase in fuel costs these houses proved to be very expensive to heat in winter.

A popular commercial type is about 7–10 m wide with glass at least 60 cm in width. The height to the ridge of these houses is about 4 m. For the amateur gardener there are some excellent metal houses constructed in exactly the same way as the efficient commercial structures. These houses require very little maintenance and are easy to manage.

When glasshouses are erected with 50–60 cm wide glass

Fig. 13 Boiler installation for heating 1 hectare of glass by hot water

Fig. 14 Warm air heater inside glasshouse with clear polythene ducting to distribute heat uniformly.

orientation on the site is of some importance. At latitudes near 50° winter sunlight at noon strikes houses at an angle of about 15° with the ground. If a house runs lengthways from north to south the glazing bars can produce a venetian blind effect inside the house which will reduce or completely cut out direct sunlight as the sun moves around the sky. On the other hand with the house running from east to west the sun will enter the south-facing side with little obstruction for most of the day.

In modern houses built in 7-m span blocks, the question of orientation may not be critical. If a choice can be made it may be preferable to arrange for the rows of plants to run from north to south in order to prevent one row from casting too much shade upon the next.

Temperature

A glasshouse must be equipped with a heating system that will maintain interior temperatures of 15–22°C whatever the outside temperature. It must also be possible to reduce the inside temperature when on hot days it tends to rise above 27°C.

Two methods of heating glasshouses are in common use:

(a) Hot water or steam circulation
(b) Warm air

Fig. 15 Peat bags set out for planting—three plants per bag.

The rate of heat loss from a glasshouse has been calculated to be of the order of 85% when ventilators are closed. From experimental work on heat transmission a coefficient has been derived which permits calculation of the amount of heat required to maintain a given temperature when the outside temperature is known. The value of this coefficient is 6·8 kgcals (kilogram calories) per square metre of effective glass area per hour per degree Centigrade. Let us take an example: suppose the outside temperature is at freezing point, i.e. 0°C, and the temperature required in the house is 15·5°C, and the total area of the glass in roof and walls is 650 m^2 then the heat demand would be:

$$6 \cdot 8 \times 650 \times 15 \cdot 5 = 68510 \text{ kgcals/hour}$$

A heating system with this capacity would have to be installed in order to ensure that the glasshouse could be used effectively for tomato production.

The cost of hot-water and steam installations is very high and many growers are turning to warm air space-heaters. These heaters usually stand in the glasshouse and burn a light oil to heat a rapidly moving volume of air which is blown through plastic ducts placed at ground level. Hot air systems are excellent for smaller glasshouses and are very flexible in use.

Automatic control of the heating installation is highly desirable. Thermostats should be mounted inside a box fitted with a small fan that draws air past the thermostat in a stream which is more representative of the glasshouse atmosphere than if the thermostat were hanging in what could be a static air pocket.

Ventilation

There are at least two reasons why air in a glasshouse must be changed frequently:

(1) Solar radiation quickly raises the temperature of the air, which on a hot day could become so high that damage to plants would occur. This hot air must be removed and replaced by cooler air to keep the temperature down to acceptable limits.

(2) The process of evaporation combined with transpiration causes the humidity of the air surrounding the plants to reach saturation. If this air is not removed the transpiration process would slow down and reduce the rate of absorption of nutrient chemicals.

For these reasons it is necessary to provide glasshouses with an efficient means of extracting the air and this normally takes the form of flap ventilators in the ridge of the house. Continuous ridge ventilators should open well above the horizontal. The area of ventilator opening should be at least one-sixth of the base area of the house. The potential loss of crop due to excessive day temperatures renders it essential that provision of ventilation should receive as much consideration as is given to heating. Automatic operation of ventilation equipment is strongly recommended for the commercial type of glasshouse. Thermostats actuate these control systems and the grower can feel confident that sudden changes in weather conditions will not catch him unawares.

Carbon Dioxide CO_2

It has long been known that the concentration of CO_2 in the glasshouse atmosphere is of first importance in the efficient operation of the photosynthetic process. Experimental work indicated that enrichment of atmosphere by two or three-fold produced a marked increase in tomato yield, but it was only in the last decade that serious efforts were made to introduce artificial CO_2 into commercial houses. Nowadays enrichment with CO_2 is accepted as a prominent feature of the tomato manufacturing process; equipment for the purpose is as important a part of glasshouse installation as is the heating system. The cost of operating a suitable CO_2 generating apparatus is more than offset by the handsome cash return.

Under normal conditions the atmosphere contains about 0·03% CO_2 or 300 ppm by volume. A glasshouse full of plants will rapidly reduce this level of CO_2 especially if the ventilation is at a minimum as so often happens during the winter. At such times photosynthesis can be quickly limited by a shortage of CO_2. The rate of carbon assimilation can be restored by introducing supplementary CO_2. It has been found that carbohydrate production is increased progressively up to a concentration 10 times the normal level, provided other factors are not limiting. In practice the accepted level of enrichment is 3–4 times the atmospheric concentration; this means that CO_2 generating apparatus is required to produce a CO_2 level of 1000 ppm.

The gas may be obtained from cylinders of liquid, from solid CO_2 known as 'dry ice', or by combustion of hydrocarbon fuels such as propane, butane or paraffin. By far the cheapest method is the combustion of paraffin which is now generally available with such a low sulphur content that it is harmless to plant life. Paraffin CO_2 generators are available in various sizes which can be controlled automatically to give CO_2 levels up to 1500 ppm.

Rooting Medium

The function of roots is to seek out water and chemicals, to absorb them and send them up the stem to leaves and other growing parts of the plant. It is vital that rapid growth of new roots should proceed continuously in order to meet the increasing demands made by the aerial portion of the plant.

A growing root possesses four distinct areas; the root cap, the meristemic or cell division region, a region where cells differentiate or become specialised, and a region where cells elongate. New cells are produced in the meristematic area just behind the root cap; further back these cells elongate and in doing so push the root cap through the growing medium. In the region of differentiation the cells become adapted for different duties. Some of the cells find themselves part of the vascular or water-conducting system, others form sugar translocating tubes, or develop into root hairs.

Cell division requires a considerable amount of energy, which can only come from the respiratory process. Absorption of nutrients takes place in the elongating area of the root and for this energy is necessary. This energy is generated by root respira-

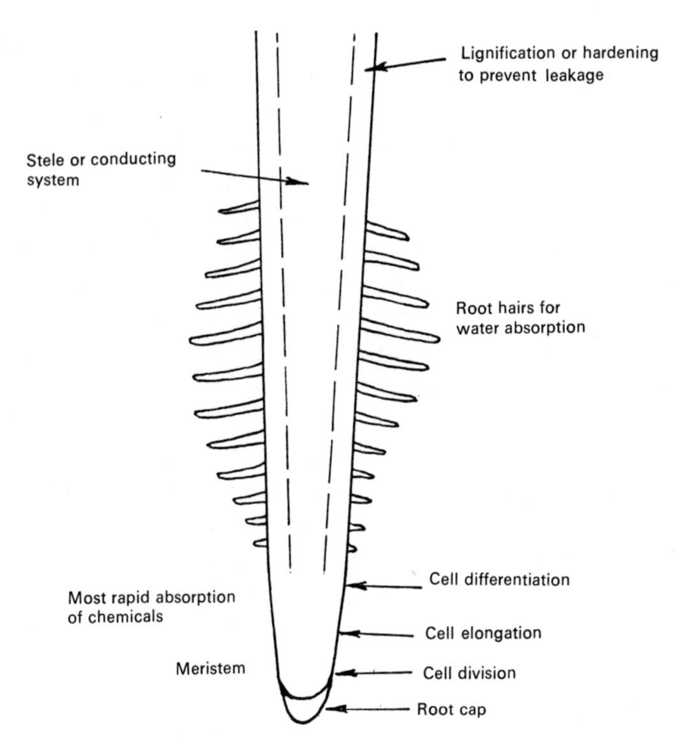

Fig. 16 Development of root.

tion which in turn requires a supply of carbohydrates transported down to the roots from the leaves. Carbon from the sugars combines with oxygen taken in from the air in the surrounding medium with the production of CO_2 and the release of chemical energy.

Water is taken up by the plant through very fine root hairs the development of which is influenced by the texture of the growing medium, its pH, nutrient content, aeration and temperature. The tomato plant will only produce the maximum number of root hairs if the soil is well aerated and has an adequate water content. Excessive dryness or high concentrations of salts inhibit root hair growth.

We see that there are certain aspects of the rooting medium that are indispensable for optimum root growth. Priority should be given to aeration since without oxygen roots will be unable to perform their tasks of water and nutrient absorption. This priority is emphasised because some growers are inclined to be preoccupied with the water and nutrient status of their soils and overlook the all-important air supply. It is quite practical to grow plants with their roots in water containing all the necessary chemicals, but unless the solution is continuously aerated the roots will die. A method of growing plants in nutrient solutions (hydroponics) will one day be developed on a practical scale. Already the difficulties of aeration and monitoring of the nutrient status of the solution are less formidable since the introduction of nutrient film techniques.

It is much easier to ensure adequate aeration if roots are able to grow in an aggregated medium. Inert solid particles break up the environment and lead to a distribution of air and solution which is congenial to roots. Sand, gravel, etc., make good aggregates and most of the experimental work on plant nutrition is carried out on plants grown in this type of medium. Many growers use inert aggregates for commercial production of tomatoes.

Peat Culture

This is a form of aggregate culture which is being employed on an increasing scale by professional growers. Granulated peat is the growing medium and it has one advantage over an inert material like sand in that it has a remarkable capacity for absorbing water but at the same time remaining well aerated. Roots find the material very much to their liking and grow profusely.

In practice the peat is mixed with a small dose of chemicals and bagged in polythene 'pillows'. These are laid in rows on sheets of plastic on the floor of the glasshouse. Young plants are planted in holes cut in the bags and roots penetrate into the peat where they are watered and fed for the rest of their life. For this type of culture it is advisable, if not necessary, to set up an automatic irrigation system since the roots are entirely dependent on the water and nutrients supplied by the grower.

Peat culture has made it possible for commercial growers to cut down the change-over time from one crop to the next to a minimum. The risk of pollution by soil-borne pests is almost negligible provided the plastic remains unbroken. These ad-advantages are not enjoyed without a price, both financial and practical. As in hydroponics, this method of growing demands a constant watch on solution composition and the operation of the irrigation apparatus.

Soil as a Rooting Medium

The most familiar and commonly used rooting medium is, of course, mother earth, but with all its good points it does have one or two drawbacks.

Soil is a very complex medium made up of components which differ widely in their chemical and physical characteristics. It is a mixture of rock particles, chemical salts, remains of plant and animal matter and live organisms. The spaces between the soil particles may be filled with either air or a solution of chemicals, known, appropriately enough, as the soil solution. The soil atmosphere which occupies those pores not containing soil solution has a higher proportion of CO_2 than the air above the soil; a product of root respiration.

In contrast with inert aggregates the great virtue of soil is its capacity to hold considerable quantities of water and nutrient chemicals. This ability to act as a reservoir varies greatly with different soil types the texture of which is determined by the size of the particles. Soil can be subjected to mechanical analysis using methods of filtration, sieving and sedimentation during the course of which five size fractions may be extracted as follows:

Gravel	above 2 mm
Coarse sand	2–0·2 mm
Fine sand	0·2–0·02 mm
Silt	0·02–0·002 mm
Clay	below 0·002 mm

Soils vary considerably in the proportion of each of these fractions. Two examples are:

Sandy loam	Coarse sand	66%	by weight
	Fine sand	18%	
	Silt	6%	
	Clay	8%	
Clay loam	Coarse sand	27%	
	Fine sand	30%	
	Silt	20%	
	Clay	19%	

Although the texture and behaviour of soil is dependent on the particle size of the soil components, it is the spaces or pores between the particles that are the decisive factors. If particles are very small they will be tightly packed together and cause the inter-particle spaces to be extremely fine. The smallest particles enclose minute channels called *capillaries*, while the space between the larger particles is known as *non-capillary* space. There is no sharp dividing line between the two types of pore space but the behaviour of the soil with regard to water retention depends on the ratio between the two volumes.

Pore space is filled with water and gaseous atmosphere. It has already been emphasised that one of the most important requirements in a rooting medium is a plentiful supply of air so that vigorous respiration might take place in the roots. Water drains quickly through non-capillary channels which are then filled with air. Capillary pores cannot be drained by gravity but only by suction; the finer the pores the greater must be the suction force to remove the water.

What happens when a soil is watered either by rain or by an irrigation system in a glasshouse? For a short time all the pore space will be filled with water—it is waterlogged and devoid of air and in no state to support root growth. Under the influence of gravity water drains down through the large non-capillary channels which fill with air drawn in from the surface. If for some reason this gravitational water does not drain away quickly enough the roots fail to work and may be seriously injured owing to lack of oxygen. Thus gravitational water is of no value to a plant and may even be harmful.

1 Portable paraffin CO_2 producer for small glasshouses.

2 Paraffin CO_2 producer with ducting to several houses.

3 Plants at harvesting stage. The good overhead leaf cover prevents sun scorch of fruit. Polythene film ducting expanded with warm air lies between plant rows.

4 Fruit discolouration due to unfavourable osmotic conditions within the plant. Broken tissues in the green areas can be seen in the cut sections.

5 The disastrous effect of grey mould *Botrytis cinerea.* This common fungus affects all parts of the plant and can spread rapidly through a crop.

6 Magnesium deficiency symptoms. Leaves have yellow areas between veins which extend until the whole leaf becomes pale yellow and useless.

7 Tomato mosaic virus. Leaves are stunted and distorted; the green colour becomes mottled and fruit shows bronze patches.

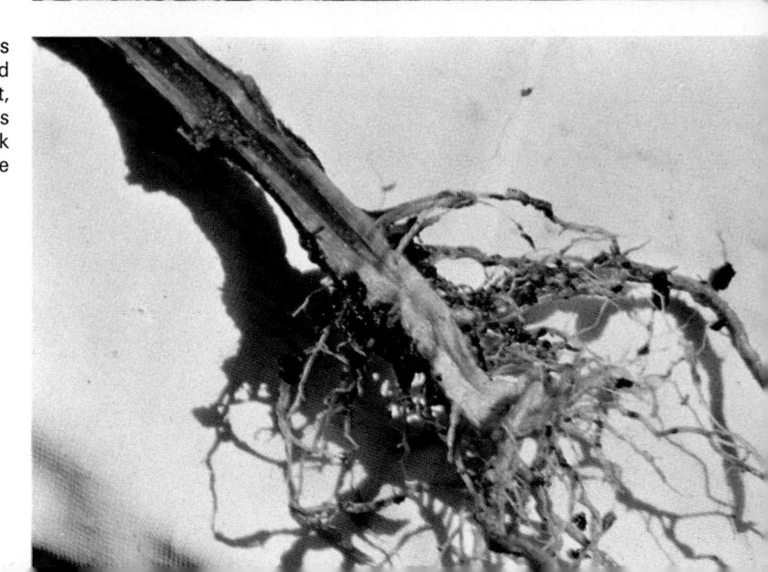

8 Fusarium wilt. Leaves turn bright yellow and die. When stem is cut, the effect of the fungus is seen as a dark discolouration of the vascular tissue.

9 Didymella stem rot. Photograph shows a magnified stem lesion. The dark specs are spore cases.

10 Leaf mould or mildev

11 Red spider – a ba leaf attack and yello spots on the fruit.

When the water has drained from non-capillary pores the soil is said to be at *field capacity*. The water remaining is held in the capillaries and as films around soil particles. At field capacity we have a pore space containing a mixture of air and water in which roots can grow freely.

As absorption of water proceeds the easily available water is used up first. The water left in the finer capillaries is held so firmly that root suction may not be sufficient to overcome the force holding the water. The moisture content of the soil at this stage is known as the *wilting percentage*, because leaves of plants begin to wilt due to the inability of roots to supply enough water to meet the demands of transpiration.

Field capacity and wilting percentage vary according to the type of soil and the available water capacity of a soil may be calculated from a knowledge of these two values. For example suppose we wish to find the quantity of available water in a cubic metre of soil with a field capacity of 45% and wilting percentage of 17%. The difference between these two is 28% which represents the available water. These percentages are based on weight, so it is necessary to know the weight of a cubic metre of the soil, for instance, 770 kg; 28% of 770 is 215 kg, which is the weight of 215 litres of water. This determination can be of value in estimating irrigation requirements under practical conditions.

The capacity of a soil for retaining water will depend on the relative amounts of sand and the silt and clay particles. For maximum aeration a high proportion of coarse particles is desirable. On the other hand, a very sandy soil has a low field capacity and would need frequent irrigation.

A loam which contains a balanced proportion of sand and clay to provide both good aeration and water-holding properties is not often obtainable, especially on sites which are otherwise favourable for tomato culture. It is therefore necessary to modify the existing soil to produce the desired conditions. Work originally carried out by the John Innes Institute shows the way this can be done. A good clay loam is the basis of a compost. Additions of coarse sand and peat are made to increase the proportion of non-capillary pore space. The effect of modifying a loam in this way may be shown by the following analyses:

	Clay loam	*J. I. Compost*
Coarse sand	27%	46%
Fine sand	30	18
Silt	20	11
Clay	19	14

It will be seen that the proportion of coarse sand has been increased considerably thus diluting the finer fractions.

If a good clay loam is available, a grower cannot do better than make a compost according to the John Innes formula of 7 parts of loam, 3 peat and 2 sand. In sandy soils peat should be added until a good balance is obtained between aeration and water retention. It is always better to err on the side of air rather than water, since it is easier to water an open soil frequently than to give enough air to a close fine-textured soil. The oxygen content of a well-aerated fertile soil is still not high enough for optimum growth of tomato plants.

The most valuable mineral fraction in a soil is the clay, the microscopic particles of which confer special properties on the soil by reason of their small size. Clay is an example of what is known as a *colloid*, a material composed of particles whose size ranges from microscopic down to the near molecular. Colloidal particles are charged with negative electricity and consequently exhibit special properties. In contact with solutions they have the power of attracting and holding on to positively charged ions. Thus, in the soil, when surrounded by soil solution, clay attracts the positive ions K^+, Ca^{++}, H^+, Mg^{++}, etc. These ions are held quite firmly and cannot be leached by water. In this way the clay acts as a reserve of minerals available for plant nutrition.

Organic Matter

The decaying remains of plants and animals in the soil are what is known as the organic matter or *humus*. It is this constituent of soil that gives it properties not found in artificial aggregates. One could debate the question whether organic humus is necessary or even advantageous for plant growth, but as soil contains the material it is advisable to understand something of its characteristics.

Humus consists of a mixture of mineral salts and complex organic molecules containing a high proportion of carbon and nitrogen. Bacteria and fungi attack these organic materials with the eventual production of CO_2 and simple chemicals that can be used by roots.

The chief value of organic matter in soils is the effect it has on soil structure. Granulation of soil—that is the binding together of very small particles into larger aggregates—is the result of the action of the fungi and bacteria which feed on the organic matter and in the process evolve gum-like substances which act as cements binding the particles together. This granulation assists in aerating the soil.

Organic matter also has a considerable capacity for retaining water because of its colloidal properties. A mixture of equal amounts of peat (dehydrated organic remains) and sand hold about 800% more water than sand alone. Like clay it can also take part in ionic exchange with the soil solution; charged organic particles attract and hold on to positive ions such as potassium and calcium.

Chemical Status of the Soil

In order to assess the value of the soil as a rooting medium we must learn a little about the amounts of the various essential elements that are present.

As we have seen, it is only the ions in solution and those attached to clay or organic colloid particles that can be absorbed by roots. These are called exchangeable ions and the elements as available nutrients. Large quantities of calcium potassium or phosphorus may be present in the soil in the form of solid minerals, but in this form they are useless for feeding plants. These minerals serve as a reservoir to replenish the soil solution if it becomes depleted.

It has been found that determinations of available or exchangeable ions give a better indication of the nutrient status of a soil than the total amounts of minerals present. Modern methods of soil analysis are based on this concept.

NPK

These letters are often used when discussing nutrition and are recognised abbreviations (the chemical symbols) for nitrogen, phosphorus and potassium, the three principal plant minerals. In what form do these materials and other chemicals necessary for plants occur in the soil?

Nitrogen

Nitrogen compounds in the soil are mainly complex proteins derived from plant or animal material. These are in various stages of decomposition. It is only when decomposition has resulted in the production of nitrate (NO_3^-) or ammonium (NH_4^+) ions that the nitrogen becomes available to the roots. The actual concentration of NO_3^- ions in the soil is extremely variable and usually quite small. Production of nitrate goes on continuously, but because the ions are not attracted to colloidal particles, they are liable to be lost by heavy irrigation. This is one of the reasons why the application of nitrogen to soils is so important in the grower's feeding programme.

Phosphorus

Phosphorus is present in the soil in combination with hydrogen, calcium and oxygen in the phosphate series of salts. Some minerals containing iron and aluminium combined with phosphorus also occur. All these phosphates are very insoluble in water and therefore relatively unavailable to plants. The three calcium phosphates decrease in solubility as the proportion of calcium increases. Thus mono-calcium phosphate $Ca(H_2PO_4)_2$ is fairly soluble in water, while tricalcium phosphate $Ca_3(PO_4)_2$ is practically insoluble. Di-calcium phosphate ($CaHPO_4$) comes somewhere in between the other two.

It is in the form of $H_2PO_4^-$ ions that phosphorus is taken up by roots. The concentration of these ions in the soil solution is therefore of importance and the pH of the soil has a great influence on the ionisation of calcium phosphate. As the pH rises from 7 to 8 the concentration of $H_2PO_4^-$ ions drops rapidly and may become too low for plant requirements.

Potassium

Certain minerals occurring naturally contain potassium which is released gradually in ionic form to be taken up by clay particles as exchangeable ions. As these ions are removed from the clay or soil solution more are released by the minerals. Sometimes the ionic exchange may be too slow for the requirements of a quickly growing crop; this explains the need for a high potassium feeding rate by tomato plants.

Calcium

The most common source of calcium is calcium carbonate (lime) but calcium sulphate is also present particularly in old glasshouse soils. Although these substances are not very soluble they are able to maintain sufficient Ca^{++} ions for plant needs.

Magnesium

This element occurs in many silicate minerals which manage to provide a sufficient concentration of Mg^{++} ions in the soil solution and on clay particles. There is firm evidence that a high concentration of potassium depresses the availability of magnesium ions and may lead to the appearance of deficiency symptoms in plants.

Manganese

One of the so-called trace elements, manganese occurs in many rocks and soils. The availability of manganese is influenced to a marked extent by the soil pH. Low pH may produce manganese in toxic quantities. In contrast, chalky soils with a high pH may suffer from manganese deficiency. Steam sterilisation may sometimes cause a dangerous increase in the manganese concentration.

. . . . .

Under the intensive cultural conditions practised in glasshouses the release of nutrient elements is not quick enough to supply the needs of rapidly growing plants. It is therefore necessary to supplement the natural resources of the soil with chemicals manufactured in factories and called fertilisers. The composition of soil solutions in glasshouses varies enormously, depending on the frequency of irrigation, manuring and the demands of the growing plants. Two properties of soil which have a pronounced effect on nutrient availability are acidity and the total salt concentration as measured by electrical conductivity.

Soil pH

The pH of glasshouse soils ranges from 5 to 8·5. Although tomato plants will grow over a wide range of pH, best results are obtained at a pH of 6–7.

At high pH (low acidity) the solubility of iron, manganese and phosphate is much reduced and deficiency symptoms may appear in the plants. Chalky soils may have a pH of 8·4 at which level the concentration of these essential elements is too low for plant requirements. Similar results can be produced by over-liming, particularly on sandy soils. The pH of a soil can be reduced by increasing the CO_2 content of the soil atmosphere. In practice this is not easy to do on well-aerated soils.

Low pH (high acidity) may cause some chemicals to dissolve to such a degree that they become poisonous to roots. Iron, aluminium and manganese—all elements required only in trace amounts—are examples. Excess aluminium interferes with phosphate uptake by accumulating in the roots. Calcium may become insoluble at low pH and addition of lime (calcium carbonate) will raise the pH. On medium to heavy soils, over-liming causes no trouble, but care should be exercised when applying dressings to light sandy soils. In these cases calcium carbonate should be applied in small amounts at frequent intervals.

Soil Salt Concentration

Soluble salts are continuously being added to soils in irrigation water and fertilisers. In the course of time the concentration of salts may reach levels that produce such a high osmotic pressure in the soil solution that a reduction in yield and, eventually, serious restriction of growth results.

It is advisable therefore to know the state of the soil in this respect. Measurements of the electrical conductivity of soil extracts provide a means of assessing the concentration of salts. Unfortunately laboratory methods have not been universally standardised and care must be taken when comparing results from different sources. Some horticulture laboratories use a saturated solution of calcium sulphate for extracting soil solutions and when conductivity measurements are made on these extracts, values of 2700–2800 micromhos are found to be suitable for tomatoes. Conductivities above 3000 micromhos are considered to be too high. (The unit of conductivity, the 'mho', is a 'reciprocal ohm'.)

Hygiene

Before commencing propagation or growing operations in any glasshouse a grower should make certain that it and all the materials used are free from insect pests and harmful organisms. Everything should be disinfected; structure, tools, equipment and the growing medium.

Soil Sterilisation

When crops are grown in the same soil year after year the tendency for diseases and predatory micro-organisms to multiply is greatly increased. Modern insecticides control pests with considerable efficiency but it is more difficult to eradicate some of the diseases which are carried in the soil. The wise grower will not attempt to grow a crop in soil which has not been treated with some form of chemical cleanser or by steam sterilisation.

Sterilisation is probably not quite the right word to use in connection with the soil since the intention is not to render the soil sterile in the sense of destroying all bacteria. Soil sterilisation in the horticultural use of the term implies freeing the soil from harmful pests and disease.

The effectiveness of chemical sterilisation is limited by the degree of penetration obtainable under practical conditions. A number of very efficient chemicals having specific functions are available. Metham sodium and methyl-iso-thio-cyanide have the same active principle which is absorbed by the soil particles. A very long period must elapse before the soil can be regarded as safe for planting. The well-known tear gas chloropicrin and a number of similar chemicals have been found to possess effective sterilising properties. These are unpleasant materials to use and must be handled with great care.

The chief drawback in the use of all chemicals is the difficulty of wetting every particle of soil. Unless every lump has been penetrated there is the risk of pockets of unsterilised soil which are a potential source of re-infection as soon as the effects of the chemical have passed. Thorough cultivation with the production of a fine tilth is a pre-requisite to efficient sterilisation.

Steam sterilisation is still the most effective method. Exposure to moist heat for ten minutes kills most of the soil-borne organisms that cause trouble. Lethal temperatures are given below:

Temperature (°C)	*Organism*
55	Weeds, earthworms, eelworm
70	Wireworms, fungi, nitrifying and other bacteria
85	Nearly all viruses
95	Tomato mosaic virus

In practice the use of steam means that the temperature of the soil must reach 100°C, which provides a reasonable margin to allow for the penetration of heat into lumps of soil.

The method of steam sterilisation practised in commercial glasshouses is based on passing steam through a number of pierced pipes buried under the soil. The area of soil to be treated must be related to the output from the boiler and to calculate the relationship between these two factors it is necessary to know something about the way steam heats the soil.

Heat is measured in *calories*; one calorie being the amount of heat required to raise the temperature of one gram of water one degree Centigrade. As this unit is very small the more practical unit for large-scale purposes is the kilogram calorie, which of course is 1000 times the calorie.

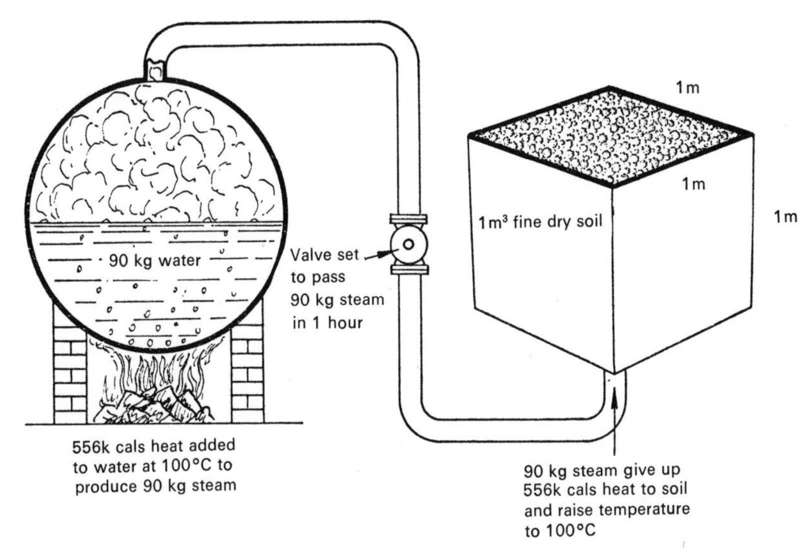

Fig. 17 Basic facts of steam sterilisation: rate of steaming is 90 kg steam/m²/hour; 90 kg steam required for every cubic metre of dry soil. These are average values suitable for practical caclulations.

If 1 kg of water at freezing point (O°C) is heated to boiling point (100°C) then it contains 100 kgcals of heat. Addition of more heat to this water will not raise the temperature but will cause the water to evaporate into steam. When evaporation is complete it will be found that a further 536 kgcals of heat have been supplied. This 536 kgcals is called the *latent heat* of steam. When the steam condenses back to water it gives up this latent heat to whatever surface happens to be present—for instance soil particles. Thus every kilogram of steam will give up 536 kgcals of heat to the surrounding soil and in doing so will raise its temperature. The pressure of the steam makes very little difference to the amount of heat obtainable on condensation. At a pressure of $5 \cdot 6$ kg/cm^2, 1 kg of steam will give 556 kgcals of heat. For practical purposes we may say that each kilogram of steam has about 550 kgcals of heat available for sterilisation.

Experiments have shown that when steam is passed into soil it condenses on the particles and gives up its latent heat at a rate dependent on the porosity of the soil and the size of its particles. Efficient heating cannot take place at a rate greater than about 90 kg/m^2/hour. If passed into soil faster than this, the steam will blow through and out into the air without heating the soil. In practice this means that a boiler with an output (actual as distinct from rated) of 500 kg per hour should steam an area of $5 \cdot 5$ m^2.

Another experimental fact is that a volume of 1 m^3 of dry soil requires about 90 kg of steam to raise its temperature to 100°C. The operative word here is 'dry'. Water requires about five times as much heat as soil to produce the same increase in temperature. Thus a wet soil will need a great deal more steam for sterilisation than a dry soil.

In commercial glasshouses two methods of steaming are in common use. In one system the steam is passed downwards from the surface and in the other steam is carried to pipes buried underground and passes through holes into the soil.

In the first method the soil may be covered with plastic sheet or metal pans and steam is blown under the covering. If plastic sheet is used it will 'balloon' as the steam fills the space and it is necessary to keep the edges from lifting. The usual way to do this is to lay chains along the edge. When the 'balloon' is uniform over the whole area the time of sterilisation can then be measured. The area covered must be balanced against the output from the boiler as calculated above.

Many growers prefer to use steaming pipes or grids buried under the soil. It is not necessary to steam trenches more than about 55 cm wide and 40 cm deep. The steaming pipes are placed in the bottom of these shallow trenches which are then covered with soil and finally by a plastic cover. The number of pipes used is determined by the boiler output. Assuming trenches are 55 cm wide and the pipes are 2 m long then a boiler producing 500 kg of steam per hour will steam five pipes at a time.

Soil should be well tilled and dry for steaming. The time required for sterilisation depends a great deal on the moisture content of the soil, but under dry conditions, 90 kg of steam are required for each cubic metre of soil. The volume of soil covering an area of $5 \cdot 5$ m^2 and steamed by 5 pipes at a depth of 30 cm with a further 15 cm loosened under the pipes would be $2 \cdot 5$ m^3. The quantity of steam required is $2 \cdot 5 \times 90 = 225$ kg. The boiler output for this area of soil must be 500 kg per hour and so the time to steam the whole plot to a depth of 45 cm would be $\frac{60}{500} \times 225 = 27$ minutes. In order to be sure that the whole of the surface layer is properly sterilised, steaming is often continued for another 10 minutes.

Repeated steaming tends to modify the crumb structure of the soil because heat speeds up the decomposition of organic matter. It is therefore customary to make additions of humus after steaming.

Sterilisation of seedling composts is almost a thing of the past now that non-soil composts are generally available at reasonable cost.

The small grower or gardener will be well advised to use peat composts for the propagation of young plants even if they are planted out in soil at a later stage.

5 Starting the Plant

Growers with profitable tomato marketing as their primary aim know that commercial success will only come from placing tomatoes on the market as early as possible in the season. The home gardener, who is not concerned with a financial profit, is usually content to take his crop during the summer. His crop is of short duration—about two months—and this is nearly always because the gardener is not able to give the close attention to some of the important details that are necessary for producing a heavy crop. If a gardener has a good glasshouse, the provision of one or two pieces of automatic equipment to control irrigation, heating and ventilation would be of immense benefit.

Early cropping necessitates sowing seed in the dark days of winter when climatic conditions are anything but favourable. The first two or three months in the life of the plant are extremely important. During this period the first fruit trusses are initiated and developed and every effort should be made to ensure that conditions most favourable to growth are maintained.

In the past few years a remarkable change has taken place in tomato production resulting in a contraction in growing time. At one time, under the most congenial conditions of high light, the interval between sowing and harvesting was about 14 weeks. In contrast with this spring-time culture, the growing time during winter was about 20–23 weeks. Nowadays the winter growing time from sowing to harvesting can be as short as 13 seeks. This startling performance can be attributed in large measure to the use of artificial carbon dioxide to enrich the glasshouse atmosphere.

Whether seedlings are started at the beginning or end of winter, the practical methods are the same. The first requirement is a compost in which roots find plenty of air, moisture and enough plant food to sustain the young plant. Peat composts provide these essentials and are obtainable in two forms; one is a sowing compost which is mixture of fine peat and sand. The seedling spends no more than a few days in this compost before transplanting and it contains only a trace of nutrient. The potting compost has the same ingredients but contains a higher proportion of plant nutrients because the young plants will be occupying this medium for some weeks.

The grower who wishes to make up his own compost based based on soil must realise that it is difficult to offer a formula without knowledge of the type of soil to be used. The well-known John Innes composts are formulated on the basis of a clay loam to which peat and coarse sand are added in approximately the following proportions:

7 parts by volume of loam
3 parts by volume of peat
2 parts by volume of sand

This formula must be modified if the loam used does not conform to the standard of clay and organic content. The correct amount of peat and sand to be added can only be determined by trials. Light loams require less peat and sand than the quantities recommended in the John Innes formula.

It goes without saying that the loam used for sowing or planting must be sterilised. Because steaming a loam containing much decomposing organic matter may lead to the formation of toxic substances the process must not take longer than necessary. Some growers like to sterilise their propagating soil well ahead of the time it is to be used to ensure the dissipation of these harmful products.

Having prepared the compost in the correct physical condition its chemical composition must be adjusted. The pH should if possible be about 6·5. If the loam is very acid then an addition of chalk up to 650 g/m^3 should be made.

It is usually assumed that young plants require plenty of phosphorus. This may be explained by the fact that is necessary for cell division and the development of new tissue. This encourages the production of a large root system which will serve the plant when water demands are considerable. An addition of 1 kg of superphosphate per cubic metre of compost is the general dose.

Nitrogen and potassium requirements may be met by 600 g of nitrate of potash per cubic metre of compost.

Seed is sown in trays and germinated at a temperature of 21°C. Some growers sow at a regular spacing of about 2 cm using frames with small holes at the required distances through which the seeds are pushed on to the compost beneath. This careful sowing has an advantage when seedlings are kept for more than a week after germination before they are transplanted. It also affords a means of assessing the percentage germination factor of any particular variety.

When seedlings are to be pricked off as soon as the seed leaves have opened there seems little point going to the trouble of spacing them so carefully. The most usual method therefore is to broadcast the seeds thinly in the box. A thin covering of sieved compost is placed over the seed and the whole moistened through a fine rose. Boxes are placed on staging and covered with plastic film. If the roll of film is wide enough the film will hang over the sides of the trays down to the ground and thus enclose heat from heating pipes that may be underneath.

As soon as the first leaves appear above the surface the covering must be removed otherwise the seedling will become drawn and 'leggy'. In a further three or four days the seedling leaves will have opened fully and they are ready for transplanting.

From transplanting to placing in their final fruiting quarters—a period of some six weeks—the young plants will be contained in a small volume of compost in a pot. These pots may be of bitumen impregnated paper about 10 cm in diameter, which need not be removed when planting out since the roots penetrate the paper.

When pricked out into the pots the young plants should be watered in to assist the root in making contact with the compost. From this stage on it is beneficial to commence enrichment with

CO_2. The gas is turned on daily, either by hand or automatically, at 8 a.m. and kept on until 3 p.m.

The seedling now enters a most sensitive stage when light and temperature are far-reaching in their effect. As soon as the seed leaves (cotyledons) expand the growing point begins to initiate leaves. This happens before the first true leaves can be seen; for a number of days from cotyledon expansion the plant is in a vegetative phase during which a succession of leaf initials is produced at the growing point. This phase ends with the initiation of the first flower truss. During the vegetative phase a high mean day and night temperature results in the production of a high number of leaves before the first truss and a low mean temperature results in the production of a low number of leaves. If an excessively high number of leaves is formed, flowering may be delayed, but within the normal range of propagating temperatures a rise of a few degrees is unlikely to cause a serious increase in leaf number. It would however induce a faster rate of growth and consequently earlier flowering. The length of the vegetative phase will vary according to the light and temperature conditions after cotyledon expansion.

When the first flower truss is being initiated a reduction in temperature will give rise to a large truss.

It will be appreciated that low temperatures favour few leaves and large trusses but that growth rate is also slow. In practice it is necessary to arrange for temperatures to result in the best balance between these two effects. A satisfactory balance would appear to be obtained at a mean temperature of 15·5°C. On sunny days the temperature may be allowed to rise to 18°C followed by a night temperature of 15°C. Dull days would require a temperature of not more than 15·5°C with a drop at night to 13°C.

During the propagation period watering is a time-consuming operation. Opinions differ as to whether spraying with a fine rose is better than watering each pot individually. When the seedlings are very small spraying is probably the better way; it is certainly quicker. However, when young plants begin to spread their leaves over the pots then individual watering is preferable.

6 A Balanced Plant

The propagating stage should produce a plant having seven or eight leaves and with the first fruit truss visible in the growing point. Nutrition of the young plants in their pots will have been carried out from the appearance of the fifth by applying a dilute solution of chemicals. Details of chemical nutrition will be given in the next chapter.

When the plant leaves begin to touch their neighbours the time has come to move them into their final quarters. This is the time for the grower to be continually aware of the physiological changes going on in the plant, knowing the environmental conditions required for optimum rate of development, and how to secure them. Let us therefore look at these various stages of growth in more detail.

Plants grow by a process of cell division. A short distance from the growing point cells enlarge by absorption of water and thus add to the size of the plants. The term *vegetative growth* is applied to this increase in stem length, leaf and root production.

After a period of vegetative growth some of the cells in the growing point pass over to the initiation of flowers which eventually lead to fruit production—*generative growth*. Having formed the flower truss the plant returns to another session of vegetative growth. This alternation continues throughout the life of the plant. How is this succession of events regulated?

Environment has a marked effect on these changes but control within the plant is exerted by specific chemicals known as growth regulators or *hormones*. Many of these substances have been isolated, analysed and found to be organic chemicals which can be prepared in the laboratory. The plant synthesises these chemi-

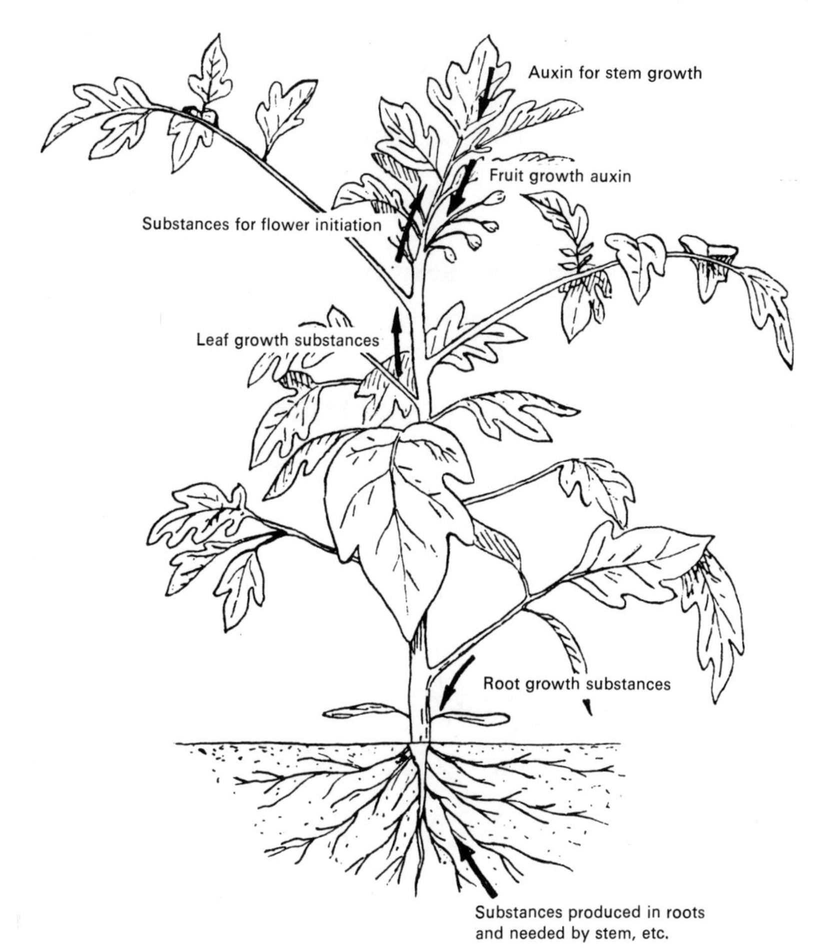

Fig. 18 Translocation of hormones for growth regulation.

cals in one part of the plant, then transports them to the area where they are to be used. For instance, mature leaves produce hormones which are required to stimulate the growth of new leaves and roots. Stem growth is regulated by a hormone known as *auxin*, made in the growing point and moved to the elongating cells a little further down from the point. The change from vegetative cell division to the initiation of flower-producing cells is governed by a chemical synthesised in the leaves and translocated to the apical bud.

Once again we see that the mysterious changes which manifest themselves as growing leaves, stems, flowers, roots and fruit are the result of elaborate chemical reactions occurring in the plant.

The effect of light, temperature and humidity on chemical reactions has already been stressed and these hormone activities are no exception. One very familiar example of these influences

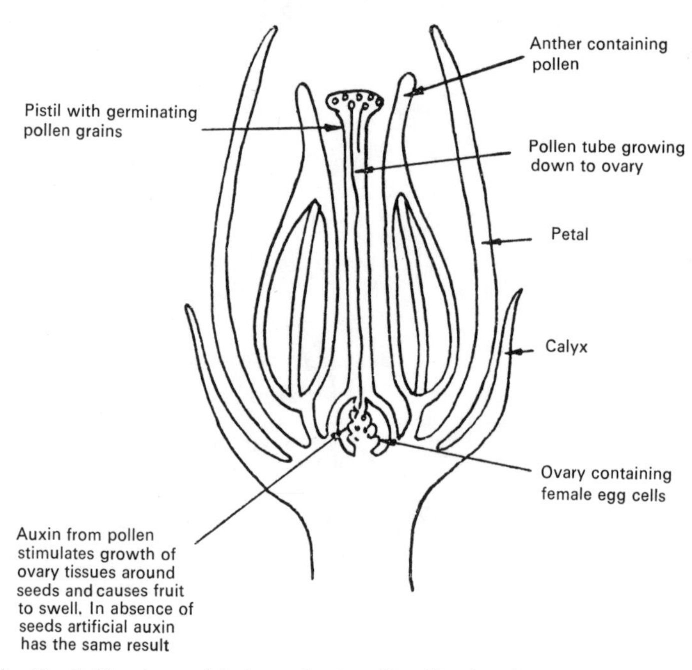

Fig. 19 Pollination and fruit production. If pollination does not occur growth ceases and the ovary drops from the plant.

is the bending of plants towards the light. The hormone auxin is inhibited by light energy so that if one side of a plant receives more light than the other that side will produce less auxin. The opposite side receiving the greater supply of auxin will then grow quicker and push the plant over in the direction of the light. The growth of fruit depends on a hormone which is present in the seeds of the fruit. Under appropriate conditions flowers will open and pollen will be transferred from the anthers to the stigma. There it will germinate and send pollen tubes down the style to the ovary where male cells from the pollen unite with female egg cells and produce seeds. At the same time hormone produced in the pollen is carried down to the ovary where it stimulates growth of tissues around the seeds. This causes the ovary to swell rapidly into a fruit. Without pollination and in the absence of the hormone the flower drops and no fruit is formed.

Hormones act by causing a concentration of sugars in the fast-growing tissues of flower buds, seed and fruit. When natural hormone had been isolated it was found to be a chemical known as indole-acetic acid. A number of chemicals related to indole-acetic acid have been prepared and are capable of acting in the same way as natural hormone. Application of these artificial hormones to growing parts of plants will produce the same sort of growth as natural hormones. For instance, the chemical napthoxyacetic acid when applied to the pistil of a tomato flower will cause the ovary to swell in the absence of pollination. Fruits formed under these circumstances have no seeds but are normal in other respects.

Since the materials for fruit development consist mainly of carbohydrates it is essential that a good supply must be available for the purpose. Application of artificial hormone will not produce good quality fruit under continuously poor light conditions. When light intensity is high as in summer, carbohydrates are abundant and it is possible that natural hormone may be a limiting factor and swelling of fruit would suffer.

Light and Growth

The effect of varying the intensity and quantity of radiation on plant development may be quite dramatic. Low light intensity produces thin, pale green stems and leaves with a low carbohydrate content. Elongation of stems is slow and roots are small because of the shortage of sugars. Fruit trusses and yield are seriously reduced in size. Thus both vegetative and generative growth suffer under low light conditions.

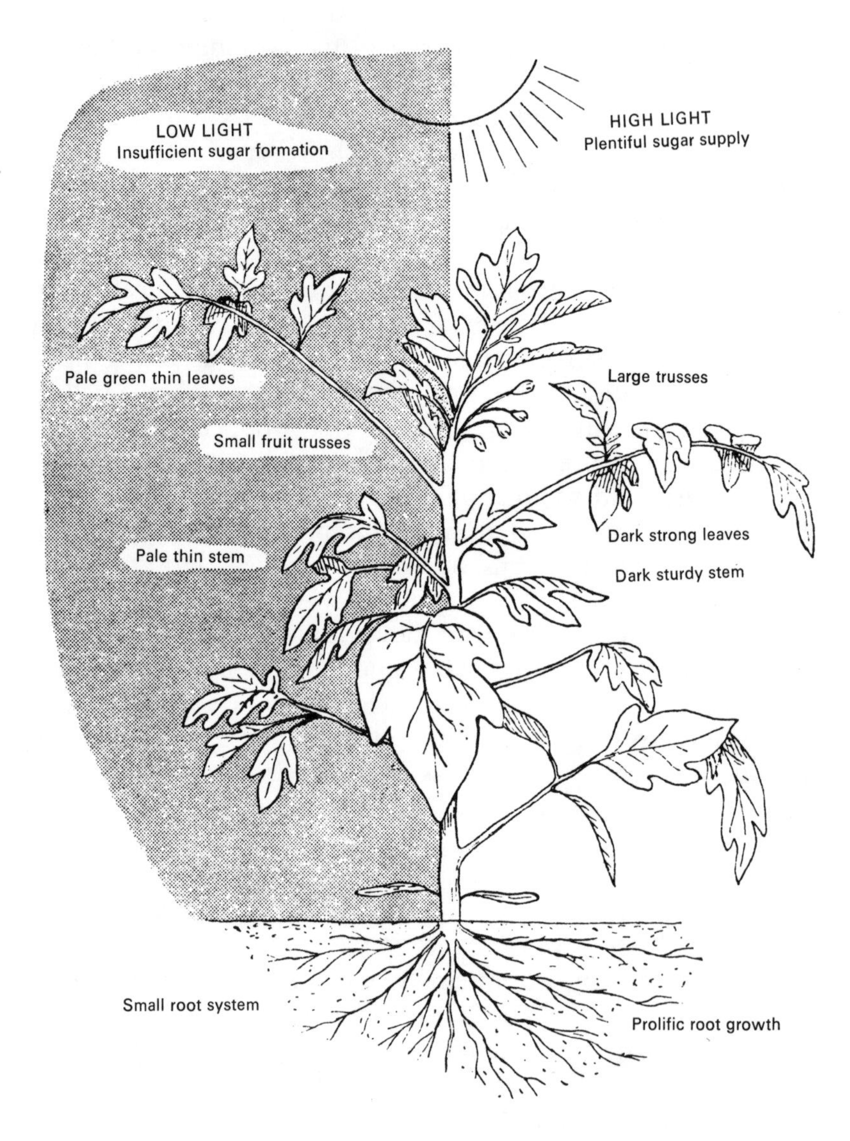

Fig. 20 Influence of light on growth.

Temperature and Growth

As most chemical reactions proceed faster at higher temperatures so plant growth is quicker under warmer conditions. Stems elongate faster but because of the greater spread of available carbohydrates the stem is thinner. At higher temperatures leaves are thin and pale in colour and the root system is small. Thus vegetative growth is more rapid but the dry weight is low. Fruit development is quick but the yield is reduced.

Low temperatures mean slower growth and because sugar loss by respiration is reduced there is a greater accumulation of sugars in the leaves and fruit trusses.

A reasonably high temperature is best for photosynthesis in good light but if the night temperature is too high then respiratory loss of carbohydrates may be too great for optimum growth. Night temperatures should be about 5°C lower than day temperatures. If day light is poor then day temperatures may have to be reduced to 15·5°C.

Enrichment of the glasshouse atmosphere with CO_2 enables the grower to maintain rather higher temperatures than those just given. For example a minimum day temperature of 20°C may be maintained even on dull days.

When flowers begin to open on trusses a new factor makes its appearance, namely the influence of temperature on pollination. The time that elapses between pollination and fertilisation of the ovary may be two or three days. Germination of pollen and speed of growth of the pollen tube down the style to the ovary depend on the temperature. If it is too low then germination may not take place, or if it does, then the pollen tube growth may be too slow to reach the egg cell in time for fertilisation. It has been found that below 14°C fertilisation and fruit set are much reduced. At 11°C it is only 25% of the maximun.

Thus the temperature for maximum tomato production is limited by the lowest temperature for fruit set. Night temperatures at setting time should therefore not fall below 14°C but to be more certain of pollination and good fruit quality the temperature should preferably be 16–17°C.

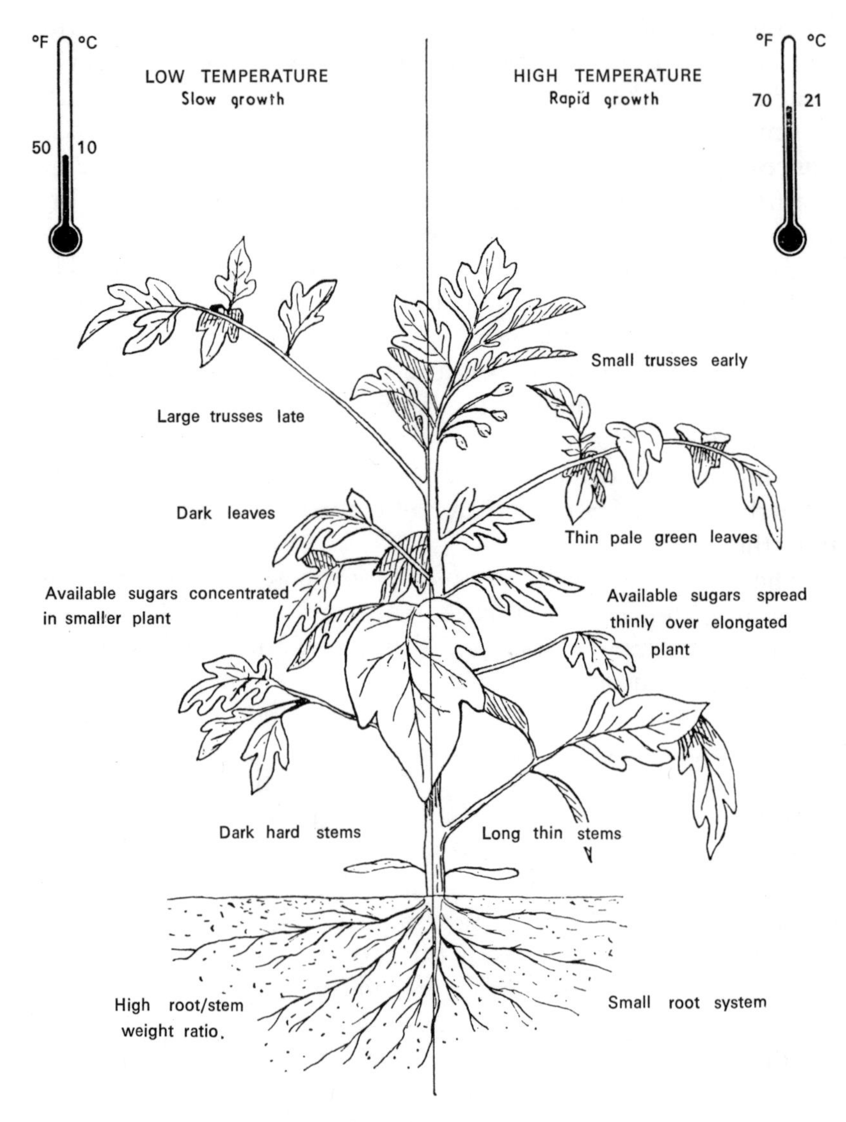

Fig. 21 Influence of temperature on growth.

Other Factors Affecting Growth

In considering the relationship between temperature and light it has been assumed that other environmental factors such as soil conditions, water and nutrient supply are acceptable.

Any interruption or deficiency in water supply causes a disturbance in the osmotic relations between cells and consequently the means by which water passes around to the various parts of the plant. The effect is to restrict growth of stem and leaves and to produce a plant grown under very cool conditions. The very high proportion of carbohydrates in these plants ensures good quality fruit although the yield will be reduced. Continued shortage of water will tend to harden the stem which will then be unable to transport the considerable quantities of water required in the later stages of growth.

A high concentration of salts in the soil solution will result in growth symptoms resembling those of a plant grown in a dry soil, that is, with the balance in favour of reproductive growth.

The relative proportion of nitrogen and potassium in the soil has a marked effect on the type of growth. Nitrates are absorbed from the soil and are used in the synthesis of nitrogenous compounds in the roots. If the supply is short then the root development proceeds at the expense of the top growth. On the other hand if nitrate is abundant then more is available for translocation to stem and leaves and vegetative growth is vigorous.

A high proportion of potassium has always been advised in tomato culture. It is not unlikely that the heavy dressings of potassium given to soils in the past were necessary because of the rather heavy nitrogenous manuring that was the common practice. When the nitrogen content of the soil or rooting medium is low there is not the same necessity for a high potassium application and the supply of these two nutrients can be kept to the 2 : 1 (K_2O : N) ratio that is satisfactory for most of the season.

7 Water and Chemical Supply

Sunlight provides plants with energy necessary for the conversion of carbon, hydrogen and oxygen into tissue-building substances. The grower has little or no control over this light except to ensure that whatever is available is used to maximum effect.

The provision of water and mineral chemicals are entirely under the control of the glasshouse grower. For this reason irrigation and chemical nutrition tended to overshadow in the minds of many growers the more important temperature factor as a means of growth regulation.

The tomato plant is very tolerant to variations in growing conditions, but there must be no plant that suffers more at the hands of its cultivators in this respect. Excessive restriction of water supply, either by accident or design, overdosing with fertilisers or continual changes in the balance of nutrients, are all tried in an endeavour to make plants do what is often impossible under the particular climatic conditions prevailing at the time. The result of these drastic control attempts is usually an unbalanced growth with consequent reduction in crop yield.

This does not mean that plants should be allowed completely free access to water and chemicals at all times. The efficient and planned regulation of water and chemical supplies constitutes one of the most important duties of the grower and there are certain rules governing these operations which should be observed. Before we go on to discuss the use of control measures in practice it will be helpful to examine the physiological factors involved in the uptake of water and chemicals.

Transpiration

Water is the life blood of the plant. It is the medium in which all reactions proceed, chemicals dissolve and are moved from place to place. It is a vital constituent of the photosynthetic process providing the element hydrogen. Most of the water taken up by the roots is lost again in the process of *transpiration*. This is a consequence of the structure of the plant and particularly the leaves.

Water in the inter-cellular space of leaves dissolves carbon dioxide and becomes involved in photosynthesis. Much of this water evaporates into the air and is carried out of the leaf pores to the surrounding atmosphere. This loss is made good by entry of more water from the vascular system in the stem, which in turn draws up water from the roots. The pull created by evaporation from the leaves is thus transferred to the water in the soil and absorption takes place through the permeable root cells.

Transpiration is a result of the difference in pressure between the water vapour in the leaves and that of the outside air. The amount of water vapour in the air is usually well below that required for saturation at normal temperatures although the artificially produced conditions in glasshouses sometimes lead to quite high humidities. During the day when temperatures are in the region of 22°C, the relative humidity may be as low as 50%. Leaf temperatures are usually a few degrees above that of the air and as the air spaces within the leaves are saturated with water vapour rapid evaporation ensues. Under these conditions transpiration may be reduced by raising the humidity of the glasshouse atmosphere; this can be done by filling the house with a fine spray and also by lowering the temperature a few degrees.

The action of light on leaves is two-fold. Apart from the energy used in photosynthesis, most of the energy is converted into heat. This raises the temperature of leaves and increases the rate of transpiration. An increase in leaf temperature from 20 to 30°C due to sunshine can cause a five-fold increase in the rate of transpiration. There is some slight compensation in that the cooling effect of evaporation prevents leaves from rising to temperatures that could be injurious.

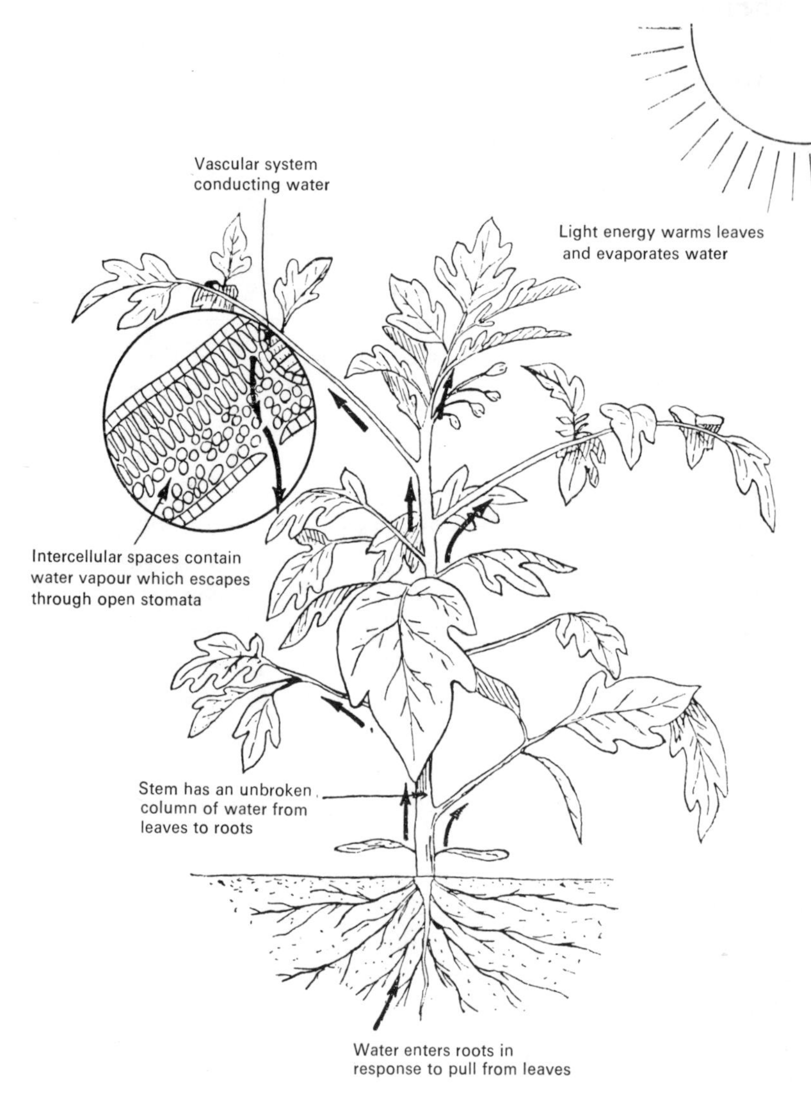

Fig. 22 Transpiration—Evaporation through stomata places the stream of water in the vascular system under tension and when transmitted to the roots causes increased absorption from the soil.

The pores or *stomata* through which water is evaporated are so constructed that they open and close according to certain outside influences. Light is the most important regulator of stomata opening. Under very dark conditions stomata may not be fully open and transpiration will be much reduced. Bright sunshine will open pores fully and the higher temperature will result in rapid transpiration.

Absorption

Water can be taken up by roots as a result of osmotic forces when transpiration pull is negligible. The amount absorbed in this way is very small and could not possibly meet the demands of a plant.

A rapidly transpiring plant requires a considerable quantity of water; on a bright day this could reach 2 litres. It is vital that the root system should be capable of providing this large quantity of water quickly enough.

The permeability of cells to the passage of water is affected by several factors of which aeration is one of the most important. A plentiful supply of oxygen is necessary for the maintenance of this permeability. Poor aeration has the effect of increasing the CO_2 content of the soil atmosphere and this is fatal for cell growth.

Moisture in the soil must be easily available if transpiration demands are to be satisfied. As the soil dries out the force required to extract the moisture increases rapidly. The suction force may be so great that the roots are unable to absorb water quickly enough to keep leaf cells turgid and the plant begins to wilt. Lack of turgidity also causes stomata to close and this stops photosynthesis. The plant ceases to function.

A high salt concentration in the soil solution will mean that it has a high osmotic pressure and this has the same effect on water absorption as a low moisture content.

We have so far been dealing with water uptake but the absorption of plant nutrients is also affected by poor aeration and high salt concentration. The concentration of cell sap is very much higher than that of the soil solution under normal conditions. Intake of salts must therefore proceed against a concentration gradient and can only do so at the expense of energy. The vigorous respiration of root cells necessary to provide this energy requires a plentiful supply of carbohydrates and oxygen. Plant nutrients are carried to various parts of a plant in the transpiration stream and if the leaves are allowed to wilt through lack of water it may be several days before normal photosynthesis and nutrient uptake are resumed.

Mineral Nutrition

The supply of mineral nutrient chemicals to plants growing in soil differs from that in peat or similar soilless composts. The soil has a reserve of chemicals which can act as a buffer against deficiencies and the practical operation of a manuring programme is a fairly simple matter. In soilless composts there is no such reserve and feeding is a continuous process demanding the use of automatic irrigation equipment on any but the smallest holding.

As the tomato requires a definite quantity of mineral nutrients

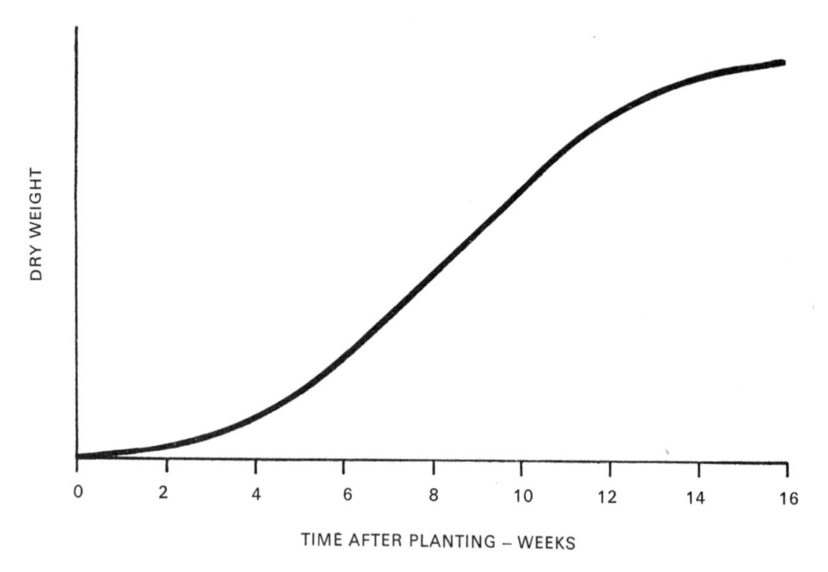

Fig. 23 A typical S-shaped growth curve in which the dry weight of a plant is plotted against time.

to produce a specified weight the requirements of soil-grown and soilless crops need not be considered separately. Therefore, if we consider a crop grown in a completely inert medium we can apply the findings to one grown in soil making an allowance for the chemicals present in the soil as determined by chemical analysis.

Two questions arise concerning the supply of mineral nutrients to be supplied by the grower. How much chemical is required and what is the pattern of uptake over the life of the plant?

The rate of increase in dry weight of a plant is not uniform throughout the whole period of growth. When plants are very small they require very little nutrient. As they grow the demand increases slowly at first and then at a more rapid rate. When the plant reaches a height of about 1·5 m, the first fruit trusses begin to ripen and are harvested. From then on the development of the plant and fruit removal make a fairly constant demand on nutrient supply. If we draw a graph showing the relation between increase in dry weight and the time from planting we get a flat S-shaped curve which is typical of most biological growth (Fig. 23).

Measurements have been made of the uptake of the major nutrients—nitrogen, potassium and phosphorus—throughout the life of a tomato crop. About 3% of the total nutrient supply is absorbed in the first month after planting and in the next 6 months another 27% is taken up. By the end of the fourth or fifth month it was found that 90% of the total nutrients had been consumed. It will be found that these values fit on an S-shaped curve and if we know the total weight of chemicals required to grow a crop of given weight then the amounts required each week can be calculated from the curve.

Experimental work has revealed that a tomato crop of 150 tonnes/hectare absorbs approximately:

136 kg N
57 kg P_2O_5
320 kg K_2O

Assuming a planting density of 30000 plants to the hectare, then 1000 plants would need approximately

$4{\cdot}5$ kg N
$2{\cdot}0$ kg P_2O_5
$10{\cdot}0$ kg K_2O

A crop of 150 tonnes per hectare would be regarded as low by present-day commercial standards. The most advanced growers turn out crops of at least 250 tonnes per hectare from a plant density of 25000 plants per hectare or about 10 kg per plant. In order to produce yields of this magnitude the harvesting period must extend over approximately 32 weeks and the total growing time from planting out at the first flower stage to about two weeks before the end of the cropping season will be about 36 weeks.

Base Dressing

The need for adding chemicals to the soil or growing medium before planting is doubtful now that it is possible to supply nutrients in closely regulated amounts by means of an irrigation system. When using peat or other inert material as a growing medium, it is advisable to add some of the essential nutrients such as phosphorus, calcium, iron and magnesium prior to planting. Phosphorus is usually added in the form of superphosphate, while calcium is applied as calcium carbonate. The latter also serves to neutralise the tendency of peat media to become acid. The so-called trace elements boron, manganese, etc., may be obtained in the form of ready mixed powders for application to the medium.

When soil is used as the growing medium, a base dressing is rarely necessary. Professional growers often have the soil analysed before planting in order to check that there are no abnormalities in the chemical content. This is particularly advisable if a crop has been grown in the soil before and has received intensive feeding. One of the most important factors to check is the total salt concentration of the soil solution. This is tested easily bymeasuring the electrical conductivity of a soil extract. A system of indices is now in general use as a convenient method of expressing conductivity in soils. These are shown in Table 3.

Table 3 Conductivity Indices

Index	Conductivity (micromhos)
0	1900–2200
1	2210–2400
2	2410–2600
3	2610–2700
4	2710–2800
5	2810–3000
6	3010–3300
7	3310–3700
8	3710–4000
9	over 4000

A conductivity index of 4–5 is suitable for tomato soils. If the salt level is higher than this then growth restriction is likely to occur. The salinity may be reduced by leaching the soil with water before planting.

A soil analysis will always include a determination of pH which should preferably be 6·5–7·0. If the soil is too acid, i.e. with a pH below 6, then calcium carbonate should be added to neutralise the acidity. Dressings vary between 340 and 700 g/m² depending on the degree of acidity. Soils having a pH over 7 are difficult to manage and a grower with such an unfortunate medium would be well advised to leave it alone and grow in a non-soil compost.

Nutrition after Planting

Application of mineral nutrients to plants in their final fruiting quarters is linked with the supply of water. We have already seen that water supply is largely a matter of meeting the demands of transpiration which depends on solar radiation. On sunny days when transpiration is rapid it is necessary to provide considerable quantities of water. In such circumstance the supply of chemicals is a simple matter. Prolonged periods of dull weather can however make feeding more of a problem.

When plants have reached a height of about 1·5 m they have a full leaf cover and the demands for water are at a constant level. It is possible to make calculations of the potential water loss by evaporation based on average solarimeter measurements and this information is available daily from many horticulture experimental stations.

If we assume a plant population of 30000 per hectare the water requirement over a wide area of Northern Europe averages out at rates shown in Table 4.

Table 4 Average Water Requirement of a 1·5 m Crop

Month	litres/plant/week
January	0·8–1·0
February	1·0–1·7
March	1·7–3·5
April	3·5–5·0
May	5·0–6·5
June	6·5–7·0
July	7·0
August	6·5–5·0
September	5·0–3·5
October	3·5–2·5

The first figure is the approximate requirement at the beginning of the month and the amount changes gradually to the second figure at the end of the month.

With a watering programme as set out in the Table it is a simple matter to devise a feeding schedule which will supply nutrients at a rate commensurate with the requirements at each stage of development of the plants.

Growers whose glasshouses are equipped with automatic irrigation systems are able to make up and apply liquid solutions of chemicals of varying composition and concentration according to the stage of development and radiation demands. The smaller grower and the amateur gardener can achieve very good results by applying his nutrients in solution using the water quantities given in Table 4, and in doing so will be following reasonably closely the S-shaped growth curve shown earlier in Fig. 23.

Preparation of Nutrient Solutions

The nutrients nitrogen and potassium are obtained from the salts potassium nitrate, ammonium nitrate or urea. These are very soluble substances and therefore dissolve easily in water to make concentrated stock solutions. Commercial potassium nitrate contains 45% potash (K_2O) and 14% N; ammonium nitrate (NH_4NO_3) contains 35% N; urea contains 46% N.

A basic solution of potassium nitrate is made by dissolving 150 g in 1 litre of water, or 1·5 kg in 10 litres. The concentration of nitrogen and potassium in this solution is:

Nitrogen: 150 g KNO_3 yields $150 \times 0{\cdot}14 = 20$ g N per litre
Potassium: 150 g KNO_3 yields $150 \times 0{\cdot}45 = 67$ g K_2O per litre

From these figures we find that the concentration of nitrogen and potassium are 2% N and 6·7% K_2O on a weight/weight basis.

The balance of nitrogen and potassium in this stock solution is suitable for feeding early plants, but in May and June a 2 : 1 ratio of K_2O/N is preferable. To obtain this solution it is necessary to add more nitrogen to bring up its concentration to half that of the potash, i.e. $\frac{1}{2} \times 6{\cdot}7 = 3{\cdot}3\%$. The extra nitrogen is $3{\cdot}3 - 2{\cdot}0 = 1{\cdot}3\%$, or 13 g N/litre. This 13 g N may be provided by adding 37 g ammonium nitrate or 30 g urea to 1 litre of basic solution.

A still higher nitrogen concentration for use in summer, in which N = K_2O is produced by adding 120 g ammonium nitrate or 90 g urea to each litre of basic solution. Table 5 summarises the three stock solutions which should cover all requirements.

Table 5 Composition of stock solutions

Stock Solution	% N	% K_2O	
High Potash	2·0	6·7	150 g potassium nitrate/litre
Standard	3·3	6·7	37 g ammonium nitrate or 30 g urea/litre basic solution
High Nitrogen	6·7	6·7	120 g ammonium nitrate or 90 g urea/litre basic solution

Before application to the soil these solutions must be diluted to a concentration more nearly that of the soil solution, which is of the order of 800 ppm. The concentration of the basic solution at 150 g/litre is equivalent to 150000 ppm, about 190 times stronger than an average soil solution. This would be lethal to roots. Thus the stock solution must be diluted about 200 times. Diluting apparatus is available capable of providing a range of dilutions. The most common dilution rate is 1 : 200 and the concentration of nitrogen and potassium in a solution of this strength will be:

High Potash	100 ppm N	335 ppm K_2O
Standard	165 ppm N	335 ppm K_2O
High Nitrogen	335 ppm N	335 ppm K_2O

Some growers use the same strength of solution throughout the season, varying only the composition from high potash at the beginning to high nitrogen when the vegetative demands are high in the height of summer. Others vary not only the composition but the concentration. Stronger solutions—say a dilution of 1 in 100—applied in the early part of the season provide a means of giving the required amount of nutrient while at the same time restricting uptake of water.

8 The Critical Stage

One of the most striking things about a chemical factory is the general absence of noisy activity such as one finds in an engineering shop. You may see a mixer stirring a liquid, hear the hum of a pump and notice a faint indefinable 'chemical' odour, but on the whole everything is disappointingly unspectacular. It is true that the vessels may have weird shapes and the maze of pipes has a peculiar sort of impressiveness but very little seems to be happening.

Careful scrutiny will reveal a thermometer here and a pressure gauge there and occasionally someone will open a valve that will control the flow of water or chemicals into a reaction vessel. The readings registered on these thermometers and gauges are of critical importance. If they depart from the values prescribed by the laboratory chemists the whole process will suffer.

If you look into a glasshouse full of tomato plants there is an equally peaceful air of inactivity. This is not intended to be a comment on any of the workers who may be present but on the chemical processes going on in the house! The plants are busy making chemicals in their quiet way and they are certainly much more attractive in appearance than the average piece of chemical plant. A well-run glasshouse will have one or more carefully sited thermometers and the grower in charge will, if he is wise, keep his eye on them at frequent intervals and regulate the temperature according to known requirements. He must also see that the correct quantity of water and chemicals is applied at the right time if the product is to turn out according to plan.

The similarity between the two forms of chemical production is so pronounced that it is surprising to find a certain indifference among some growers to the need for precise methods of control.

The propagation stage usually lasts for five or six weeks and the temperature during this period determines the number of leaves between the trusses and the number of flowers in each. The period from planting out to flowering and setting of the third and fourth trusses is therefore one of the most critical of the season. In most districts this stage will extend from February to March, while in very early growing areas it will begin in January. The amount of light available at this time of the year is very limited; frost and snow are not uncommon and cold north-easterly winds often accompany any bright sunny periods. Not the sort of weather to make things easy for the grower anxiously watching the flowers developing on his plants.

Having produced sturdy young plants during the propagating phase the grower must be sure that the medium and space where these plants will grow to maturity are suitable in every respect. The plants will be occupying this area for many weeks, will require support and must be properly spaced out so that large leaves have plenty of room in which to perform their useful function.

If the glasshouse soil is used for planting it must be free from disease and root pests. Depth of soil is not important; in fact there may be advantages in having a shallow soil provided an automatic irrigation system is installed. The most advanced method of growing in peat bags or nutrient troughs has shown that heavy crops can be produced from plants whose roots occupy no more than one-hundredth of a cubic metre of growing medium.

Planting Systems

At one time growers were accustomed to plant in rows alongside heating pipes, with plants spaced at 30 cm apart in the rows. This gave a plant population of 35000 plants per hectare which produced a heavy early crop although the fruit size was on the small side and in the later stages of the crop the yield fell away due to overcrowding.

Experimental work on spacing has led to a gradual reduction in plant density with the result that growers now plant at 45 cm apart in the rows giving a plant population of about 32000 per hectare. In deciding how many plants are required to fill a house, a grower should prepare double rows 50 cm apart with a working path about 1 m wide between the double rows.

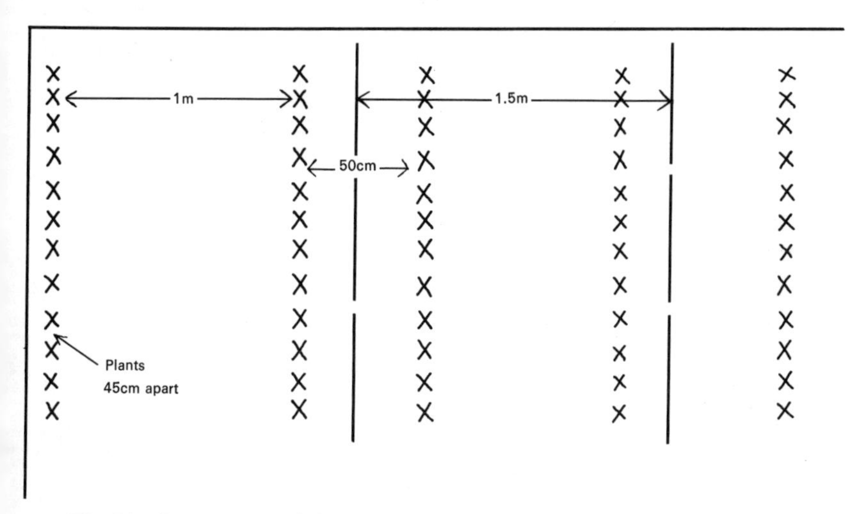

Fig. 24 Arrangement of plants in cropping house.

If very early cropping is the aim then some form of container separated from the ground should be employed. This can take the form of 25-cm pots, plastic bags or troughs, or even more permanent concrete troughs. Such troughs need be no more than 15–20 cm deep and about 30 cm wide. They are filled with either sterilised soil compost or a soilless compost.

Planting Out

Plants grown in 10–cm pots through the propagating period should be brought to a stage when the first truss breaks into flower. The balance between vegetative and reproductive growth which was a feature of the propagating period should be preserved after planting. A plant which is allowed to grow vigorously after it has made roots in the new medium will become thick stemmed

Fig. 25 Barrel type dilutor. Available in capacities from 20–160 litres. Dilution variable from 1 : 100 to 1 : 400.

with large coarse leaves. Flowers will be pale in colour and if they pollinate the resulting fruit may be misshapen and of poor quality. In contrast, a plant which is subjected to restriction in root development in one form or another has a thin woody stem and small leaves. Fruit sets quickly but the yield may be low.

Fig. 26 Aspirated thermostat for accurate measurement of glasshouse temperature. The end cover is removed to show thermostat behind which is a fan for drawing air through the apparatus. When connected to valves in the heating system, automatic control of temperature is obtained.

By planting at the flowering stage the young plants will have been under close control and have developed a growing habit which will carry over after planting out provided favourable conditions are maintained. The plants will be too big for the propagating bench and the usual procedure is to stand them in the rows prepared for them in the glasshouse. Roots penetrating through the paper pot are prevented from entering the soil or compost by placing plastic film beneath the pots.

When the first flower on the first truss is open the plant may be put into holes prepared for them in the growing medium. The moisture content of the medium should not be too great, preferably a little below field capacity. Roots will find great difficulty in penetrating into dry soil so we have a means of preventing too rapid a development of plants when they are placed in the new

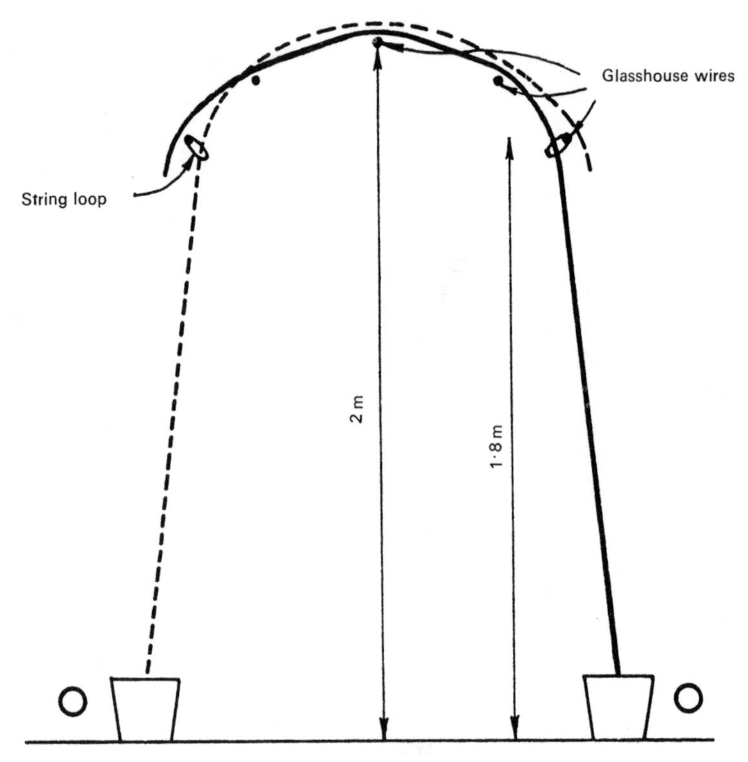

Fig. 27 Arched method of training plants.

environment. Careful irrigation after planting enables the plant roots to grow out into the surrounding medium gradually. Every watering should contain dissolved nutrients in the form of a diluted high potash solution.

Temperature Control

The relatively low temperatures at which tomato crops used to be grown were dictated by the low winter light conditions in the absence of CO_2 enrichment. Use of this addition to growing technique has rendered it possible to raise growing temperatures without creating unbalanced plants.

The low limit of temperature for tomato culture is 13°C. Below this temperature growth rate is reduced. At the other end of the scale a day temperature above 27°C will not produce increased rate of growth but may result in lack of vigour, lower yield and poor fruit quality.

Optimum temperatures for balanced growth lie between these two extremes and a great deal of experimental work has been done to find the best combination of day and night temperatures. It has been established that night temperatures should be lower than day temperatures in order to reduce the rate of respiration and allow translocation of sugars down to the roots. As a result of all the investigations, it is recommended that a minimum day temperature of 20°C should be the aim. If winter sunshine occurs during the day it will help to lift the temperature which can be allowed to rise to 24°C. It will be a rare event for winter temperatures to exceed this value but should this happen in the warmer days of spring, then the use of ventilation will keep the temperature down to the correct level. At night the temperature should drop to 16°C.

As the season progresses and light intensity improves it will be necessary to use more ventilation during the day and possibly at night in very warm weather. Photosynthesis has an optimum temperature range of 24–26·5°C under good light conditions and there is no point in allowing the temperature to rise above 26°C. At temperatures over 29°C transpiration demands may be so great that roots may fail to supply water quickly enough and the plants wilt under the strain. The importance of accurate temperature control cannot be over emphasised and reliable thermometers should be installed in every house. They should be protected from direct sunlight and preferably be in a moving stream of air.

Carbon Dioxide Enrichment

The value of enrichment of the glasshouse atmosphere with CO_2 has already been stressed. The higher temperatures now employed

Fig. 28 Tomato crop trained over wire.

during winter days are only possible because of the maintenance of a much higher volume of CO_2 in the glasshouse.

Fruit number and size are both increased by added CO_2 and the early commercial grower cannot afford to be without some means of generating the gas and introducing it into his houses. CO_2 enrichment should commence at sunrise when plants begin photosynthesis and should continue until sunset when light intensity becomes a limiting factor. When it becomes necessary to ventilate in order to control temperature the use of CO_2 is uneconomic.

Moisture Control

Water which is absorbed by the plant and not lost by evaporation causes elongation of cells and general vegetative growth. This occurs at night and under low light conditions and plants then tend to charge themselves with water and become soft and succulent. Some control over this can be exerted by increasing the concentration of the nutrient solution in the irrigation system. This will have the effect of increasing the osmotic pressure of the soil solution and make it more difficult for the roots to absorb moisture.

Transpiration is rapid when the atmospheric humidity is low, a condition which prevails when ventilators are open and light is intense. Relief may be given to plants by spraying them with water and thus increasing the humidity around the leaves.

Nutrient Balance

It has already been shown how the relative amounts of nitrogen and potassium in the soil can exert a marked effect on the balance between vegetative and generative growth. Nitrogen in excess causes an increase in the size of cells but gives them thin walls. This makes the leaves large and succulent and susceptible to attack by insects and fungi.

Potassium does not enter into the structure of plant tissue but plays an important role in a number of chemical processes within the plant. It enters into photochemical reactions and the osmotic relationships between cells. The large leaves produced by excess nitrogen and low potassium are relatively inefficient photosynthesisers and this reduces the carbohydrate content. A high concentration of potassium in the nutrient solution corrects the ill effect of too much nitrogen. There is evidence that a high potassium content in the soil improves the quality of fruit both in colour and flavour.

Fruit Setting

When, as a result of maintaining the appropriate growing conditions, a balanced plant is produced there should be no problem in achieving free pollination and good setting of fruit.

For flowers to set fruit the pollination process must be unchecked. Pollen will be sterile below 12·8°C and above 32°C. A temperature between 17·8°C and 20°C has been recommended as best for the highest level of fruit setting.

Humidity is an important factor in pollination. It is sometimes difficult to maintain the humidity level during warm dry weather and the spraying, which is useful for reducing transpiration, is also beneficial for pollination. If a large number of pollen grains fertilise an equal number of egg cells then the number of seed produced is also large and the result is a solid, well-formed fruit. When pollen is scarce or does not germinate, the number of seeds in the fruit is small and the fruit is badly misshapen.

9 Full Production

The normal time for development of fruit from setting to maturity is about eight weeks. As flower trusses form at the rate of one every week or so, by the time the fruit on the first truss is ready to ripen there will be about six or seven trusses of fruit swelling on the plant.

It is most important that plants should receive no serious check during this period. The feeding regime suggested in Chapter 7, combined as it is with a steadily mounting irrigation application, should take care of the essential requirements of water and mineral chemicals. But growers are human and may be caught on the wrong foot, or their soil conditions may not be all that could be desired, so we should consider the effects of a failure in water or nutrient uptake.

Transpiration usually wins in the competition for the limited amount of water available in such emergency circumstances and the new rapidly growing areas are deprived of water. Thus, young trusses near the head of the plant may dry up; the knuckle behind the calyx of the flowers turns yellow and the flower bud drops before coming into bloom.

It is not uncommon to see flowers dropping after pollination is assumed to have taken place. This happens even when light and temperature are favourable and pollen is plentiful. Evidence seems to indicate that the pollen has failed to germinate through lack of water. Partial germination may produce only one or two seeds resulting in small fruits that never swell to full size. Little can be done about 'cut' or dropped bloom but they can be recognised as a signal that the plant is not taking up water at a sufficiently rapid rate.

Serious water shortage has more far-reaching effects. The irresistible demands of transpiration may have to be met by withdrawal of water from the fruit, with a consequent breakdown in the wall tissues. This manifests itself particularly at the base of the fruit furthest from the calyx. As the fruit swells the tissues which have been destroyed by lack of water turn brown and rot. They are then easily invaded by disease organisms. Tomatoes in this state suffer from what is known as 'blossom end rot'.

The most common cause of water shortage, apart from a failure of the human or automatic irrigation equipment, is an inadequate root system. This may be due to poor soil conditions or the action of root pests and disease. It is unfortunate also that the plant chooses this period in its development to shed a proportion of its root system. The lush, vegetative type of plant produced by high temperatures in the absence of CO_2 enrichment or by over-watering, has large leaves and small roots, making it particularly prone to suffer the ill-effects of sudden drought. On the other hand if control of growth has been too severe in the early stages there is a likelihood that the stem will have become hard and 'woody'. The vascular tissues of the stem will have been partially broken down, thus reducing its capacity to transport water.

When plants suffer from water shortage due to a sparse root system, relief may be given by repeated overhead damping. Some growers cover the surface of the soil with a mulch of peat or straw to reduce evaporation from the surface and to induce new roots to grow from the base of the stem.

Ripening of Fruit

Ripening of fruit is a process which is influenced by temperature and not, as some think, by light. Provided they are kept at the same temperature tomatoes will ripen at the same rate whether in darkness or in light.

The colour of tomatoes is largely determined by the ratio of two chemicals, lycopene (red) and carotene (yellow). At temperatures above 27°C lycopene is not produced and fruits develop a yellow colour. A temperature of 24°C seems to be the optimum for good red colouring. If tomatoes are exposed to direct sunlight or are in warm spots in the glasshouse they may remain at over 27°C for most of the day and thus turn yellow. At night when the temperature drops lycopene will be produced and a red tinge will begin to develop. If this is continued for the few days required to reach maximum colour it will be found that the fruit will never acquire the full red colour, but will turn to an indeterminate orange and may be patchy. Fruit coloured to picking stage under these conditions takes one or two days longer than that ripened below 27°C and is therefore more mature and softer and less fitted for eating or sending to market. We have already seen that 24–27°C is the optimum temperature for photosynthesis, so that both from this point of view and that of ripening there is no advantage in allowing glasshouse temperatures to rise above 27°C. Every effort should be made by ventilating, and control of heating, to prevent temperatures from exceeding this value.

Another cause of uneven or 'blotchy' ripening is the breakdown in the tissue of the fruit wall attributed to withdrawal of water from the fruit to supply unusual or sudden demands of transpiration. The outward appearance of fruit suffering from this disorder is a mixture of red and green patches. If the surface of the tomato is shaved off brown streaks of dead tissue can be seen beneath the green areas. Blotchy ripening due to this cause seems to occur most frequently on plants which are soft and watery. The explanation and practical means of prevention centre on the osmotic relationships within the plant. If the soil has a low salt concentration as a result of frequent heavy irrigation, the plant sap will also have a low overall osmotic pressure, although there will be differences from one part of the plant to another. Water moves out of the stem into various organs in response to these differences in osmotic pressure. Swelling fruits absorb water from the stem because the osmotic pressure of the sap in their tissues is higher than that of the nutrient solution in the vascular system. Suppose now that transpiration makes sudden and heavy demands for water which cannot be met by the roots. Water in the stems will come under a tension which may be high enough to overcome the osmotic suction force drawing water into the fruit. In a soft, watery plant this osmotic tension is relatively low and the transpiration suction is very likely to reverse the flow of water and cause plasmolysis, or collapse of the fruit cells. The breakdown in tissue which follows interrupts normal development of the ripening process, and so these parts of the fruit remain green or perhaps yellow. If, however, the osmotic pressure of the cell sap is high enough there will be sufficient suction to balance or overcome the hydrostatic tension in the vascular system and flow of water out of the fruit will be prevented.

The practical way to guard against 'blotch' from this cause consists in maintaining a salt concentration in the soil high enough to 'condition' the plant. This is done by applying water judiciously and keeping up a good supply of nutrient chemicals. The feeding programme suggested in a previous chapter aims to create the soil moisture and nutrient conditions required at each stage of plant growth.

The conductivity index of 4 (Table 3) is suitable for plant nutrition and should produce osmotic conditions that will preserve fruit quality. Should the index fall below 4 there will be a tendency for uneven ripening to occur. A salt concentration giving an index much higher than 5 may be enough to reduce yield by producing small fruit although the colour may be excellent.

Associated with blotchy ripening is a disorder known as greenback, so-called because the skin on the shoulders of the fruit around the calyx remains green when the rest of the fruit is red. Sometimes the green turns yellow when the fruit is fully ripe. As with the other forms of discolouration there appear to be two causes, one due to temperature and the other to osmotic conditions. Fruit that is exposed to hot sunshine may show this symptom, the remedy being obvious. More frequently, greenback occurs on hard plants having a cell sap with an excessive osmotic pressure. Correction of this form of greenback lies in applying more dilute nutrient solutions in order to lower the salt concentration in the soil.

The puzzled grower may wonder just where he must draw the line between strong or weak solutions, since at the one extreme he may produce greenback and at the other blotchy fruit. In practice the range between these two extremes is wide enough to be easily manageable provided watering and nutrition are carried out with reasonable precision. Carelessness will only lead to trouble.

In the early part of the season tomatoes on the middle trusses are sometimes afflicted with a malady known as hollow or 'boxy' fruit. These terms describe the symptoms for when cut open large air spaces are found in the locations which are normally filled with seeds and pulp. The cause is not known with any certainty. Faulty pollination, unbalanced nutrition, lack of water, heat and cold have all been suggested as being responsible. Some investigations have indicated that the balance in development of seed-bearing tissues and fruit wall may be upset by high temperature in the early part of the season. It would seem that slight reduction in growth rate in the early stages of development and careful attention to nutrition are the best means to guard against hollow fruit.

Nutrient Deficiencies

Among the disorders which may affect plants are those which arise from a deficiency of one or more of the mineral nutrients. Under the closely controlled growing conditions maintained in glasshouses they are comparatively rare and scarcely ever serious. Occasionally some abnormality in soil balance occurs and one of the chemical elements becomes unavailable. When this happens symptoms specific to the element concerned appear in the plants. Recognition of these deficiency symptoms is useful, but diagnosis is complicated by the fact that environmental factors may sometimes produce similar signs.

Deficiency of nitrogen is not unusual during the propagating stage, when plants are raised in small pots, and again later in the season when they are carrying a full load of fruit. There is a general lightening of the green colour and the lower leaves turn yellow. Growth will be much reduced, stems and veins in leaves develop a purple colouration due to the abnormal formation of anthocyanin pigments and fruit is a pale green when immature. Immediate correction of this condition follows an application of quick-acting chemical such as urea or ammonium nitrate. An excess of nitrogen in the soil produces a succulent vegetative type of plant which will be difficult to bring into a fruiting state.

Phosphorus deficiency may occur in the presence of phosphate in the soil if the pH is too high as a result of over-liming. Growth suffers probably because of a poor root action; leaves are dark

Fig. 29 Nutrient film technique. Plants grow in plastic troughs through which a shallow stream of nutrient solution flows continuously, emptying into a collecting tank and then being pumped back to the head of the troughs. Control apparatus can be incorporated to maintain the pH and nutrient status automatically.

green with a purple anthocyanin colouration and occasional areas of dead tissue. It is easy to confuse plants raised in cold conditions with those suffering from phosphorus deficiency. Probably the simplest way to correct this trouble is to adjust the pH of the soil. Leaf tests on tomato plants usually indicate a very high level of potassium, and deficiency of this element is so unlikely that its symptoms need not be considered.

It so happens that the heavy applications of potassium to glasshouse soils is largely responsible for the most widespread deficiency, namely that of magnesium. Plants suffering from a lack of this element may be seen in the best regulated houses. Excess of potassium appears to depress the uptake of magnesium from the soil and in these circumstances applications of magnesium sulphate to the soil offer little or no answer to the problem. Reduction of the potassium content of the nutrient solution will gradually improve matters and spraying of the leaves with a solution of magnesium sulphate has been found effective. The characteristic symptom of magnesium is a chlorosis or yellowing between the veins of the leaves.

Of the minor nutrient elements, iron and manganese may become deficient in heavily limed soils with a high pH. Yellowing of the growing head of the plant is a symptom of iron deficiency, while a lack of manganese results in a light green leaf with small yellow areas between the veins. Treatment of the soil with chemicals in which iron and manganese are 'sequestered' and protected from the alkaline reaction of the soil is the simplest way to deal with this problem.

In the foregoing catalogue of some of the troubles that may beset the grower in producing and ripening his fruit it has been evident that they all arise from some abnormality in the environment. We cannot escape the conclusion that the closest attention must be given to every aspect of this environment such as soil condition, temperature, atmospheric humidity, water and chemical supply and that precision is essential. Errors or carelessness in this respect will surely lead to trouble.

Training the Plant

A tomato plant is unable to support itself on its stem and will creep along the ground unless held up by some form of support. The usual method is to twist the stem around string tied to wires hung from the structure of the glasshouse. The 'arched' method of training extensively employed in Guernsey in the Channel Islands appears to be one of the most effective. Wires are arranged as shown in Fig. 27. The main tying wires are placed over the plants at about 2 m above soil level. If these wires are a few centimetres from the vertical in the direction of the path, then the space between the plants in adjacent rows will not become too crowded with leaves. This will improve the efficiency of the area both as regards photosynthesis and air circulation. Three wires are placed over the path at equal distances apart and rising a few centimetres towards the centre of the path. The centre wire should not be more than 2·2 m from ground level for convenience in working.

After having placed the young plants in their permanent quarters and allowed them to grow to about 35 cm in height they are ready for tying. Three-ply hemp or polypropylene string is used for this purpose. A loose loop is made around the stem below one of the lower leaves and the string is then tied to the overhead wire. As the plant grows it is twisted around the string in easy spirals. When the head of the plant has reached the level of the tying wire it is looped with string so that it passes on the path side of the wire. As the stem grows towards the centre of the path it rests on the wires without further tying. Each stem merely requires placing between those coming from the opposite side of the pathway. Occasionally a twist may be given to the stem in order to prevent an obstinate head from going upwards towards the roof and also to bring the fruit trusses below the leaves. This protects the flowers and their fruit from hot sunshine and makes fruit harvesting easier. Having reached the far side of the arch the head is encouraged to hang down by another twist to the stem, or in stubborn cases by tying it to the stem of a neighbouring plant. The effectiveness of this form of training depends to a great extent on the efficiency with which the operations of side-shooting and defoliation are carried out.

Side shoots

Tomato plants produce shoots in the axils of leaves with amazing rapidity. These should be removed as early as possible since if left to grow they will absorb energy that can be more profitably used in fruit development on the main stem. If allowed to grow unchecked, side-shoots can become a serious nuisance in what is called the trimming of plants and also a harbour for pests.

Leaves

The usefulness of leaves as producers of sugars is to some extent counter-balanced by their interference with air circulation around the plants. When plants have reached a height of about 1·5 m the leaf area can be considered to have attained maximum effectiveness from the point of view of photosynthesis and potential transpiration. If no leaves were removed and plants grew to 3 m, photosynthesis and transpiration would be no greater at the same light intensity. The reason for this is that in the taller plant more leaves would be in the shade and the area of those in the light would be no greater than that on a 1·5-m plant. One can assume therefore that as long as 1·5 m of stem carried leaves the remainder could be removed without impairing the efficiency of the plant. The disadvantage of removing all leaves in this way is that some fruit may be exposed to hot sun with consequent ill-effect on colouring. It is better to remove leaves gradually with the primary object of increasing air circulation around the whole plant. When the plant is about 1·5 m high, all leaves up to the bottom truss may be removed. At the same time one leaf may be taken out between each truss provided that, by its position, it is contributing little to the process of photosynthesis. As trusses reach ripening stage so general leaf removal progresses up the stem.

The three operations of training, leaf removal and pinching out side-shoots constitute what is usually known as trimming the plant. It is a time-consuming operation and if allowed to fall behind can result in a reduction in crop. The secret of success lies in early removal of side-shoots before they have grown to any length. Every plant should be trimmed at least once a week with the operations carried out in the following order:

(1) Leaf removal
(2) Twisting around string or placing on wires
(3) Shoot removal

Removal of shoots is left to the last so that in the event of the head of the plant being damaged during twisting a shoot may be allowed to grow in its place.

10 The Fifth Column

Chemists in charge of a manufacturing process can usually expect to carry on their work confident that their apparatus will give good service for long periods and that, under normal circumstances, there are very few people about who would deliberately interfere with their activities.

Not so the grower; he is more than likely to find his glasshouses and plants attacked by armies of saboteurs whose sole purpose appears to be to live and multiply by destroying everything in sight.

We have already described some of the physiological disorders that may affect plants, and have seen that they occur when there is some maladjustment in the various physical and chemical processes that are in progress. The cause is almost invariably environmental and appropriate remedial action is comparatively simple. It must be remembered that sometimes the symptoms of disorder appear long after the damage was done and it is too late to correct the trouble.

When external agencies in the form of insects, bacteria, fungi and viruses begin to intrude then special security measures must be introduced.

Insect Pests

Insects do not produce disease symptoms but may act as carriers. Most of the troubles caused by insects are in the form of damage to plants rendering them physically incapacitated. Of the numerous pests that would be described in a detailed entomological work we shall only consider here those that can prove seriously harmful in glasshouses.

Root-knot eelworm

These microscopic worms enter the root hairs from the soil and as they spread into the main roots cause large swellings or galls. These cause the roots to assume a distorted knotted shape. Functioning of the root system breaks down and plants are unable to take up water and nutrients to meet their needs. They become stunted and weak and, if the attack is severe enough, they may die. Worms remain in the root tissues at the end of the season and it is advisable to remove all old roots before making winter preparations for the next crop. Control by steam sterilisation and proprietary soil fumigants depends on the thoroughness with which they are carried out. The tendency of eelworms to retire to the lower strata of the soil during winter does not make the problem of eradication any easier.

Symphilids (White insects)

Very severe damage to roots can be caused by these small members of the millipede family. If a plant is lifted quickly from the soil the insects may be seen moving rapidly into the surrounding soil. The roots of the plant will be seen to be short, as though someone had been at work with a pair of clippers. Because of the continual destruction of new roots the plants become stunted and make very little growth. Symphilids also have the habit of going down into the deep layers of the soil during the winter and may therefore be out of range of steam or chemicals. Sterilisation, however, is still very effective in keeping infestation at a minimum. Further treatment of the soil, after planting out, with BHC or parathion is also recommended.

Insects which attack plants above ground level are more numerous and require continual control throughout the season if fruit yield is not to suffer. Fortunately a number of chemicals are available which when used under favourable conditions, destroy these marauders most effectively. The most economical way of using these chemicals is in the form of atomisable fluids sprayed into the houses as a fine mist from high pressure guns.

Red Spider Mite

Presence of this pest is first indicated by minute pale specks on the surface of leaves. When the leaves are turned over it is possible to see the very small red spiders which cause the specks. If no action is taken the leaves become so damaged that they wither. The spiders make webs from leaf to leaf and the plant becomes useless.

Control for red spider should begin very early in the season. A mixture of chemicals which kills both adult spiders and eggs sprayed into houses before insects are seen often does much to delay the appearance of this pest. A number of proprietary chemicals are available for fighting red spider mite and if infestation is serious it may be necessary to try a number in succession. The mites tend to become resistant to a particular chemical group and it is advisable to change to another which is unrelated chemically.

White fly

This pest is not so widespread as red spider and attacks vary in severity both geographically and from season to season. Very small moths live by sucking the sap from leaves, usually from underneath. They do not damage the foliage but secrete a sticky substance called honeydew which covers leaves and fruit. A brown mould grows over this honeydew, seriously interfering with the functioning of leaves and making the fruit dirty. Chemical controls for white fly are numerous but as with red spider mite resistance varies considerably. The most effective insecticides are probably the organo-phosphorus compounds although even these are showing signs of inadequacy. Control measures must begin early and be continued at intervals of seven to ten days. Weeds inside and near glasshouses should be removed as they are harbours for swarms of these insects.

Caterpillar

Since the introduction of DDT this pest is no longer the menace it was a few years ago. An early preventative application of DDT in smoke or aerosol form will keep crops free for many weeks.

Leaf Miner

Flies and maggots feed on leaves, producing white streaks over the surface. In a short time the whole leaf may be affected, thereby limiting the area for photosynthesis. Attacks occur at any time but the pest is usually most troublesome in the propagating stage. Steam sterilisation will deal with the insects which may be present in the soil, while control during the growing season is exercised by parathion smoke or spray.

Thrips

Usually seen at the propagating stage rather than on mature plants, these small flies produce white spots on the leaves. Damage is only slight, but it is important to eradicate the pest as soon as possible, since it is a carrier of the virus disease, spotted wilt. DDT, BHC and parathion are effective controls.

Germ Warfare

Bacteria and fungi are microscopic forms of animal and plant life, respectively. *Bacteria* enter plants through wounds or natural openings, multiply, feeding on the tissues, and produce specific symptoms within the plant. A single bacterium may produce many hundreds of millions in twenty-four hours. Some bacteria form spores which are capable of resisting the usual methods of sterilisation.

Fungi are responsible for more of the tomato diseases than bacteria. They are forms of plant life which obtain their food either from dead organic matter or from other living organisms. Fungi form spores which are seed-like in character and when conditions are suitable they germinate and produce new infections. Spores are spread by wind, insects, water and on the clothes of workers.

Damping-off and Collar Rot

These attack seedlings and young plants and are caused by fungi of the phytophthora species. They penetrate directly into the stem at soil level. Temperatures between 20 and 30°C favour the growth of these fungi and a high organic content in the soil provides a good breeding ground. The stems of young plants are softened and become so thin that they topple over and die. Control consists in using steamed compost containing a low proportion of organic matter and keeping temperatures in the region of 15·5°C.

Root Rots

These are caused by various fungi and are encouraged by poor drainage and low soil temperature at planting time. They are common diseases in soils which have not been carefully sterilised. Symptoms include blossom drop, dry setting, weak growing points and wilting in the sunshine even with a moist soil. On examination the roots have a brown skin which can be easily pulled off the white inner core.

Little can be done about plants suffering from root rot, except to reduce the demand for water by shading and overhead damping. New roots can be encouraged to break out by putting down a mulch of peat on the soil surface. At the end of the season all roots remaining in the soil should be removed and burnt. Then the soil must be sterilised very thoroughly, preferably by steam.

Verticilium and Fusarium Wilts

These are very similar in character, the former being rather more widespread. The symptoms are first, a sudden wilting of the lower leaves followed by the appearance of yellow patches on these and other leaves. The chief difference between the two fungi is in the temperature of their development; verticilium develops at 21–24°C whereas fusarium requires a higher temperature of 26–30°C.

Both fungi enter the roots and interfere with the ascent of water up the stem, thus causing wilting. Toxic substances are also produced which assist in damaging the conducting vessels and eventually bring about the death of the leaves and plant. If the stem is cut open along its length the pith will show a brown colour due to the poison.

Treatment of plants attacked by these two wilts is difficult and not permanent. Very thorough steam sterilisation of the soil will go far towards eliminating these diseases.

Didymella Stem Rot

This appears to be more common on outdoor tomatoes but it can be devastating in glasshouses. Symptoms are dark, sunken lesions on the stem which soon spread around the stem and cause death of the plant.

Infection may arise from the seed or the soil and may be carried on the hands of workers. Diseased plants should be removed carefully from the greenhouse and the soil for some distance around the site of infection should be treated with a solution of Captan.

Leaf mould

This disease is caused by the fungus *cladosporium fulvum* which thrives on the high temperatures and humid conditions in glasshouses. Pale brown or yellow spots appear on the surface of leaves, the underside of which show a brown-grey growth of fungus spores. Spread of spores is rapid, and if left uncontrolled the disease soon kills the whole plant. Cool dry conditions within the prescribed temperature regime should be maintained with good circulation of air around the plants day and night. Control can be achieved by spraying with Benomyl or dichlofluamid fungicides.

Grey Mould (Botrytis cinerea)

This is one of the most common of fungus diseases in the tomato. The fungus is present on dead and decaying matter and gains entrance through wounds on the leaves and stems where it kills tissues. The diseased area becomes covered with a brown-grey mould which can spread quickly to all parts of the plant. Cleanliness in the glasshouse is essential; decayed or dead leaves and stalks should be removed without delay and a dry atmosphere should be preserved. Fungicide sprays such as those used for leaf mould are also effective against botrytis.

Virus Diseases

In some ways the most serious of the parasites of the tomato are the viruses. Very much smaller than bacteria or fungi, viruses are chemical substances with a complex protein structure. They may be extracted from plant tissues and obtained in pure crystalline form with definite, large molecular weights. These virus compounds have the unusual capacity to multiply in the plant tissues but have no metabolism of their own. When a small quantity of plant juice containing virus is rubbed on the leaf of an uninfected plant the virus begins to act on the cells of the leaf and spread by multiplication until it is found in all parts of the plant. As new growth is produced it too becomes infected from within the plant itself. Any means by which a minute amount of virus can be introduced into a plant through a wound or abrasion serves to transmit the disease from one plant to another. Insects, handling, rubbing of workers' clothes, pruning knives, are common ways of causing infection.

Tomato Mosaic Virus (*TMV*)

One of the most infectious of the virus diseases. Mottling of leaves occurs about nine days after infection, starting with young leaves. In serious attacks the plants become stunted because of the reduction in fully effective leaf area. The infection soon spreads to flower trusses which become weak and drop their bloom. Fruit which may be swelling and reaching maturity is also affected and it colours unevenly.

The disease can spread like a plague during ordinary cultural operations unless very strict precautions are taken. The principal difficulty is to spot the infected plants early enough so that they can be isolated.

Mosaic is often introduced during the propagating stage, though it is not known how the virus makes its initial entry. One of the most effective ways of controlling mosaic is by careful handling of seedlings with special emphasis on cleanliness. Hands should be washed in soap and water before seedlings are touched and smoking should be prohibited while work is in progress. Seedlings should be closely scrutinised and any that may look at all unusual should be rigorously eliminated.

It is now possible to inoculate seedlings with mild strains of TMV which give some protection against the more severe strains that occur later in the plant's life. The mild virus is sprayed on the young seedlings with a spray gun and within a week the seedlings are protected from infection.

When TMV appears in plants that are fully grown it is too late to remove them. The only course is to keep them growing as freely as possible and to encourage them to come to terms with their affliction. This may be done by cutting down light intensity by shading, keeping the atmosphere humid and increasing the nitrogen content of the nutrient solutions.

Viruses of different composition produce symptoms of widely varying characteristics ranging from bright yellow mottling to curled twisted leaves. Names are given to these manifestations of virus infection such as aucuba mosaic, single-virus streak, spotted wilt and enation mosaic. All these virus diseases reduce the vitality of plants and may produce blemished fruit but they rarely result in death of plants. There is one, however, which can lead to serious losses. This is *double-virus streak*, which is a combination of TMV with potato virus, and is liable to occur when workers handle tomato plants after working with potatoes. Leaves become mottled and covered with brown spots and many of them wither and die. Dark brown streaks appear on the stems and growth is weak. Fruit setting fails and any fruit which is swelling becomes rough and covered with brown spots. All plants must be carefully lifted, removed from the house and burned.

11 Crop Programmes and Varieties for North European Latitudes

1 An Early Long Crop

To produce tomatoes from early March to September.

Minimum requirements

A good modern glasshouse sited in a high winter light area; a heating installation capable of maintaining a temperature of 20°C; an automatic irrigation system with liquid feeding dilutors; equipment for CO_2 enrichment.

Propagation

Sow first week November.

Prick out in 10-cm pots 12 days after germination. Space pots at 17 cm apart on the bench.

Compost: fine peat and sand or proprietary potting compost.

Temperature: 21°C for germination. After pricking out: day 20°C, night 15·5°C.

Feed with every watering after the fifth leaf has expanded.

Enrich with CO_2 from dawn to dusk.

Prepare cropping house to receive plants paying particular attention to hygiene.

Planting

Plant density: 3 plants/m². Plant in double rows at 150 cm centres and 50 cm between rows. Space plants at 45 cm apart in rows.

Move plants from the propagating house when leaves overlap on the bench; stand in planting positions on plastic film.

Plant when about half the plants have one flower open.

Continue CO_2 enrichment.

Feed and water daily or twice a day if necessary on hot sunny days.

Temperatures: day minimum 20°C, night 17°C. Ventilate at 24°C. If automatic ventilation is not available and weather is variable operate vents to maintain 21°C.

Tie plants at base and attach string to wire overhead.

Cropping period

Temperatures: day 20°C minimum; night 16·5°C.

Feed with every watering.

Training

Remove side-shoots regularly and leaves as fruit trusses clear. Twist stems around the string and when long enough lay over the wires above the path. Protect fruit from direct sunshine. Rigorous attention to trimming is necessary for the best results.

Continue CO_2 enrichment until ventilation is required.

Watch for pests and disease and take precautionary measures before attacks become serious.

2 Mid-season Crop

To produce tomatoes from April to September.

Requirements

Very similar to those for an early crop. This crop is more suitable for areas where winter light is less favourable. The heating system should be capable of maintaining a temperature of 15·5°C.

Propagation

Sow January, first week.

Temperature: 15·5°C or higher if possible.

Other operations as for early crop.

Planting

March 1st.

The slower growth due to lower temperatures may be compensated by improving light conditions.

Temperatures and feeding programmes should follow as nearly as possible those in Programme 1.

3 Late Crop

To produce tomatoes from May to September. Not recommended for market supply except as a catch crop. Although costs of production are relatively low the fruit is likely to reach the market when prices are at a minimum. The domestic gardener and the smallholder with a local market would be interested.

Requirements

A good glasshouse or plastic tunnel; heat is not necessary but is desirable; if irrigation is to be by hand then a dilutor for hose use must be fitted; an adequate ventilating system must be provided.

Propagation

Sow end of February

Use same methods as in Programme 1

Planting

Early April

Considerable benefit will be obtained if plants are put into some form of container rather than the greenhouse soil, particularly if the weather has been cold prior to planting.

As minimum temperatures will depend on outside conditions ventilation will be the only means of some form of temperature control. As the season advances temperatures should not be allowed to rise above 21°C if at all possible.

All feeding, watering and training operations as in previous programmes.

Tomato Varieties

Hundreds of tomato varieties have been produced and every seed catalogue will recommend at least 10–20 in glowing terms. The choice can be narrowed down to about 5 depending on the type of grower and his aim in growing the fruit.

Commercial growers with market demands to satisfy and production problems to face are still looking for the variety which will meet all requirements. One would have thought that the first criterion would be eating quality or flavour. Unfortunately, in some communities this is not the case; the spread of supermarkets, with their insistence on packaging, has made display counter appearance and uniformity in size of paramount importance. As a consequence, the commercial grower is too often committed to varieties which can be expected to yield only tomatoes of billiard-ball size and shape which will colour uniformly.

In addition to these requirements, the commercial grower demands a variety which has a high early yield, a vigorous plant and a resistance to disease. This is a tall order but in the last ten years plant breeders have made considerable progress towards meeting these characteristics. Most of the varieties now available are F_1 hybrids and since there is a gradual change in those which find popular favour it would be of little value to list them in a book of this nature.

The so-called 'straight' or non-hybrid varieties such as the well-known Potentate, Moneymaker or Ailsa Craig are still available and, of course, the seed is very much cheaper than that of the hybrids. The smallholder with a local market outlet and the home gardener for whom table quality is a prime consideration could do very much worse than try these well-established varieties.

Index

Acidity, 12, 13, *See also* pH
Acids, 11
Aeration of soil, 42–46, 71
Aerial environment, 33–40
Alkalis, 11
Aluminium, 5, 34, 50
Ammonium nitrate, 11, 76, 94
Anions, 9
Atomic weights, 6–7
Atoms, 6
Auxin, 61–62

BHC, 99, 101
Bacteria, 47–52, 101
Base dressing, 74
Bases, 11
Benomyl, 103
Blossom end rot, 90
Blotchy ripening, 91
Boron, 29
Botrytis cinerea, 103
Boxy fruit, 93

Calcium, 5, 29, 49
  carbonate, 50, 57 ,74
  hydroxide, 11
  phosphate, 11
  sulphate, 49, 51
Capillary pore space, 44–45
Carbohydrates, 18, 28–29, 63
Carbon, 5, 28
  compounds, 17
  dioxide, 7, 28, 39–40, 59–60, 85, 87
Carotene, 90
Caterpillars, 100
Cations, 9
Cells, 15–17, 26–27, 40–41
Cellulose, 18
Chemical compounds, 5–6
Chemical elements, 5
Chemical reactions, 12, 22, 61–62
Chemical sterilisation of soil, 52
Chemistry, 4–12
Chlorophyll, 27
Chloropicrin, 52
Cladosporium fulvum, 103
Clay, 43–46
Club root, *See* Root-knot eelworm
Coarse sand, 43–46
Cotton rot, 101
Colloids, 46
Colour of tomatoes, 90–92
Composition of tomato plant, 29–30
Composts, 45–50, 56–58
Concentration of solutions, 10, 77
Conductivity index, 50, 75, 92
Copper, 5, 29
Crop weight, 2, 74

DDT, 100, 101
Damping off, 101
Dichlofluamid, 103
Didymella, 103
Differentiation of cells, 40–41
Diffusion, 10, 15
Dilution of solutions, 10, 77
Double virus streak, 105

Electrical changes in clay, 46
  conductivity, 13–14, 74–75
  dissociation, 9
  energy, 20
  forces, 6, 9
Electrons, 9
Elements, chemical, 5–6
Energy, 18–20
Environment of plants, 25, 32–39
Evaporation, 20, 69, 75–76

Fats, 18

Fertilisation, 65
Fertilisers
  nitrogen content, 77
  potassium content, 77
  role in tomato growing, 74–78
Field capacity, 45
Fine earth, 43
Fine sand, 43, 46
Flower trusses, 59, 60, 89
Flowering temperatures, 88
Fruit, conditions for growth, 63, 65, 88
  ripening, 90–91
Fungi, 101
Fusarium wilt, 102

Generative growth, 60
Germination, 26, 58
Glasshouses, construction of, 33–37
  heating of, 37–38
  hygiene, 51
  relative humidity, 21
  transmission of light, 33
  ventilation of, 38–39
Granulation of soil, 47
Greenback, 92
Grey mould, *See Botrytis cinerea*
Growth of plants, 25–30
Guernsey training of tomato plants, 95

Heat energy, 20, 22
Hexose sugars, 18
Hollow fruit, 93
Hormones, 29, 60
Humidity, 21, 88
Humus, 46–47
Hydrogen, 5–7, 17–18, 28
Hydroponic culture, 42
Hygiene of glasshouses, 51

Indole-acetic acid, 63
Insect pests, 98
Ions, 9
Iron, 5, 27, 29, 50, 95
Irrigation, *See* Watering

Kinetic energy, 19

Latent heat, 53
Leaf miner, 101
Leaf mould, 103
Leaves, 97
Light, effect on growth, 59, 63–64
  effect on ripening, 90
  effect on transpiration, 69, 76
  energy, 22–24
  requirements of plants, 32–36
Lycopene, 90

Magnesium, 5, 29, 49, 95
  sulphate, 11
Manganese, 5, 27, 29, 49–50, 95
Meristems, 40
Mineral nutrition, 29, 71–78
Moisture level of soil, 44–45
Molecular weights, 7
Molecules, 7
Molybdenum, 30

NPK, 48
Naphthoxyacetic acid, 63
Nitrate of potash, *See* Potassium nitrate
Nitrates, 11, 48
Nitric acid, 11
Nitrogen, 5, 29
  deficiency, 93
  effect on growth, 67, 87
  in compounds, 76–77
  in humus, 47
  in soil, 48
  requirements of plants, 58, 73
Non-capillary space, 44–45
Nutrition, 29, 71–78

Oils, 18
Organic compounds, 47
Osmosis, 15
Oxygen, 5, 29
  in compounds, 7–8, 47
  in soil, 46
  requirements of plants, 28, 41, 71

pH
  defined, 12
  effect on intake of minerals, 49–50
  high, 75
  of propagating composts, 57
  of soils, 50, 75
Parathion, 99, 101
Peat, 42–43
Phosphates, 48
Phosphoric acid, 11
Phosphorus, 5, 29
  deficiency, 94
  in soil, 48
  pentoxide, 8
  requirements of plants, 57, 73

Photosynthesis, 22, 24, 27–28, 69, 85
Phytophthora fungi, 101
Planting out, 80–81
Plasmolysis, 17, 92
Pollination, 62, 65
Polysaccharides, 18
Potash, *See* Potassium oxide
Potassium, 5, 29, 49
  deficiency, 94
  effect on growth, 67, 87
  hydroxide, 11–12
  infertilisers, 77
  in soil, 49
  nitrate, 7–8, 11, 58, 76–77
  oxide, 8, 73–74
  requirements of plants, 58, 67, 73
Potato virus, 105
Potential energy, 19
Potting composts, 57–58
Pricking off, 58
Programmes for crops, 106–108
Propagation, 52–60
Proteins, 18

Raw materials for tomato production, 2
Reactions, chemical, 12, 22
Red spider mite, 100
Relative humidity, 21
Respiration, 5, 28, 40, 65
Ripening of fruit, 90–91
Root-knot eelworm, 99
Root rot, 102
Roots, described, 40
  inadequate, 90
  medium, 40
  water-absorption, 41, 71

Salts, concentration, 51, 74–75
  defined, 11
  effect on growth, 67, 75
  high concentration, 67
  in cells, 16–17, 91–92
Sand, 43–46, 57
Setting of fruit, 31, 65, 88
Side shoots, 96
Silicon, 5
Silt, 43–44, 46
Sodium, 5
Soil, analysis, 57, 74–75
  as root-medium, 43–48
Solutions, 9, 51, 77
Sowing, 58
Steam, heating of glasshouses, 37–38
  sterilisation of soil, 52–55
Stem rot, 103
Sterilisation of soil, 52–55
Stomata, 71
Sucrose, 18
Sugars, 18, 28
Sulphur, 5, 29
Sulphuric acid, 11
Superphosphate, 58, 74
Symphilids, 99

Temperature, control of, 37
  during propagation, 58–59
  effect on growth, 65–66
  for ripening, 90–91
  requirements of tomato plants, 65, 85, 88
Thermometers, 38
Thrips, 101
Tomato mosaic, 104
Trace elements, 30, 74
Training plants, 95–96
Transpiration, 30, 69–71, 75–76
Transplanting, 58
Trimming plants, 96–97
Turgidity of cells, 16–17

Urea, 76, 93

Vacuole, 16
Varieties of tomato, 108–109
Vegetative growth, 59–60
Ventilation of glasshouses, 38, 85
Verticilium wilt, 102
Virus diseases, 104
  inoculation, 105

Water, 12, 30
  assimilation by roots, 71
  effect of shortage, 67, 89–90
  in soil, 45, 71
  requirements of plants, 69, 71, 87
Watering, during propagation, 59
  effect on nitrates, 48
  in relation to sunshine, 75–76
Weeds, 52, 100
White fly, 100
White insects, See symphilids
Wilting percentage, 45
Wireworms, 52

Zinc, 5, 29